RUDIMENTS OF MINING PRACTICE

Other books published within the Series on Mining Engineering

M.L. Jeremic:
Elements of Hydraulic Coal Mine Design
1982

Donald A. Trotter:
The Lighting of Underground Mines
1982

Siegfried von Wahl:
Investment Appraisal and Economic Evalution of Mining Enterprise
1983

RUDIMENTS OF MINING PRACTICE

by

Cedric E. Gregory
BE, BA (*Adel*), BEcon, ME, PhD (*Qld*)

Emeritus Professor
University of Idaho

Visiting Professor, King Abdulaziz University
Jeddah, Saudi Arabia

First Edition
1983

TRANS TECH PUBLICATIONS

Rudiments of Mining Practice

Distributed in North America by
KARL DISTRIBUTORS
16 Bearskin Neck, Rockport, MA 01966, USA

and worldwide by

TRANS TECH PUBLICATIONS
P.O. Box 266
D-3392 Clausthal-Zellerfeld
Federal Republic of Germany

and by

TRANS TECH S. A.
CH-4711 Aedermannsdorf, Switzerland

ISBN 0-87849-042-6
ISSN 0720-6267

Printed in the United States of America

PRINTED AND BOUND IN THE UNITED STATES OF AMERICA

To miners the whole world over:
Who have contributed so very much
to our standard of living

Preface

Most mining schools at North American universities offer a sophomore course called "*Introduction to Mining Engineering*" in order to introduce mining technology to students at an early stage and to develop and maintain their professional interest at a time when they are otherwise submerged in 'dry' basic science and engineering subjects.

I have personally delivered such a course of lectures in six universities in three countries over a span of 23 years; and I am particularly impressed with the success of such an established aim to give an overview of the whole wide spectrum of mining engineering operations, necessarily in a superficial manner. This not only excites and inspires the students at this level, but it also prepares them in expectation of the content of specific detailed in-depth subjects to follow in the junior and senior years. It therefore helps them to plan their onward study programmes and enables them to engage in summer work projects to advantage. This is one particular course in which a slight amount of 'overlapping' can be fully justified.

I have also found that such a broad course structure provides an attractive elective subject for those civil engineering majors who seek ultimately to engage in civil engineering contract work, and who therefore need some formal instruction in such aspects as rock blasting as practised in tunnelling, quarrying and rock excavation work generally. The same applies to those geology and geological engineering students who because of the shortage of mining engineering graduates, and consequently to the detriment of the mining industry, are currently able to find work opportunity as 'mining engineers'; and who, without having attended such a course, have had little or no knowledge of mining engineering practice. To these people such a broadly patterned course is vital, if only as an elective subject.

Nevertheless, there is no suitable textbook available to service this type of broad shallow course offering. There are several with such titles as *"Elements of Mining"* and *"Introduction to Mining"*, but they tend to give too much space to a few topics in too much depth and omit others altogether. Otherwise, they are padded with case histories, many of which tend to legitimize obsolescent operating practices. This is very confusing for students.

Consequently, after all these years, I have been encouraged to review my notes and to prepare this manuscript, based also on my 34 years of experience as a hardrock miner, a mining engineer, a mining executive and a mining professor. I trust therefore that this little book will help mining school instructors to rationalize their curricula, to crystallize their lecture programmes, and to inspire more and more mining students, in both the older and newer mining schools of North America at least. A similar opportunity is presented to the schools of geology and earth sciences. With such a little book, there should be ample opportunity to illustrate the principles covered by quoting extemporaneous examples of the instructor's own mining experience, if any, to the class. If handled appropriately, this is where inspiration can be generated.

My apologies are offered to those instructors who have specialized in the mining of coal or industrial minerals. I admit that my emphasis (in the book) is on hardrock metalliferous mining, but this happens to be the sector in which the range of mining expertise attains its broadest scope.

For the cogent reasons expressed above, this book is not designed as a definitive treatise on mining engineering, even though it purports to cover a wide range of mining engineering operations. On the other hand, it aims purposely to embrace these features in a simplistic manner, thereby avoiding the risk of confusing these younger sophomore students with too much detail in a course to be delivered in a quarter or semester.

Yet it should be understood by all readers, including students, that mining engineering encompasses a vast industrial field of complex operations, many of which are based upon the uncertain vagaries of nature, and therefore subject to many imprecise and indeterminate specifications and other factors. As a consequence, it necessarily embraces many features of the industrial arts (as well as of the precise sciences), calling for the exercise of mature common judgement which can be derived only from hard practical 'earthy' experience, as distinct from the nebulous precepts of theoretical psychology.

Both Imperial and *Système International* units are used, and a Conversion Table is included at the end of the book. There is also a Glossary as an aid to those readers unfamiliar with mining terminology.

The book is designed primarily, but not exclusively, as a textbook. It is written in a straightforward readable manner. Therefore one would like to believe that it would have a significant appeal, not only to students, but to prospectors, to residents of mining towns, to accountants, lawyers and other professionals associated with mining company activities, to stockbrokers and investors, to visitors (even tourists) to mining centres, as well as to general readers. It should therefore contribute to a more realistic understanding by the public at large of the national importance of mining. Because of this need, it should be a reading requirement in all liberal arts colleges for students in economics, journalism, political and social sciences; and by candidates for public office in any country.

C.E.G.

Acknowledgements

The author desires to acknowledge the kind assistance of Dr. Karl H. Wolf for his pertinent suggestions in reviewing the manuscript.

Credits are also due to the following for permission to use diagrams and photographs:

Atlas Copco AB, Fig. 22.
Bucyrus-Erie Co., Figs. 18, 20.
Jeffrey Dresser, Figs. 16, 31.
John Wiley & Sons, Inc., Figs. 32, 41.
Fried. Krupp GmbH, Fig. 19.
Linden-Alimak AB, Fig. 6.
O & K Orenstein & Koppel AG, Fig. 19.
Society of Mining Engineers of AIME, Figs. 6, 26, 27, 35.
Wagner Equipment Co., Fig. 29.

Special thanks are due to Pergamon Press Inc. for permission to draw on Section I of "A Concise History of Mining" (Gregory, published 1980) for much of the material in Chapters 1, 2, 3, 6 and 7.

List of Figures

Fig. 1: A Simplified Diagram of the Rock Cycle
Fig. 2: Typical Modes of Occurrence of Veins and Lodes in the Crustal Host Rock
Fig. 3: Three Aspects of Mining Activity
Fig. 4: Four Classes of Mining Operations
Fig. 5: Timbered Shaft Set for Rectangular Shafts
Fig. 6: Alimak Unit
Fig. 7: Cross-section of Hypothetical Ore Vein
Fig. 8: Longitudinal Section of Vein
Fig. 9: Simple Wooden Chute Gate
Fig. 10: Method of Mining Ore Block by Shrinkage Stoping
Fig. 11: Round Timber Stull
Fig. 12: Method of Mining Ore Block by Cut-and-Fill Stoping
Fig. 13: Chute Timbering
Fig. 14: Hypothetical Underground Coal Mine Development
Fig. 15: Two Methods of Mining Coal Underground
Fig. 16: A Continuous Miner
Fig. 17: A Coal Strip Mine Operation
Fig. 18: A Walking Dragline Excavator
Fig. 19: A Bucket Wheel Excavator
Fig. 20: A Power Shovel
Fig. 21: Open Cut Mining of a Wide Lode
Fig. 22: Several Types of Pneumatic Rockdrills
Fig. 23: Timbered Crib for Ground Support
Fig. 24: Details of Square Set Timbers
Fig. 25: Square Set Timbering Arrangements
Fig. 26: Yieldable Steel Set Details
Fig. 27: Yieldable Steel Sets in a Drift
Fig. 28: Rockbolts
Fig. 29: A Wagner Scooptram
Fig. 30: Loading Ore Cars from Draw Points
Fig. 31: A Shuttle Car
Fig. 32: Scraper Units
Fig. 33: A Double-drum Scraper Unit Operating in a Scram Drift
Fig. 34: Various types of Ore Cars
Fig. 35: A Swedish Train
Fig. 36: Simple Hoisting Mechanisms
Fig. 37: Examples of Ordinary Lay Hoist Ropes
Fig. 38: Various Steel Wire Rope Sections
Fig. 39: A Single Rope Friction Hoist
Fig. 40: Rope Arrangement of a Friction Hoist
Fig. 41: Hoisting Cage for a Small Mine
Fig. 42: Typical Fan Selection Diagram
Fig. 43: Ducted Ventilation of a Blind Heading

Contents

	Page
1. Introduction	13
2. Mineral Deposits	23
3. Three Classes of Economic Minerals	29
4. Mineral Exploration	33
5. Sampling and Assaying	37
6. Four Classes of Mining Operations	39
7. The Scope of Mining Activity	67
8. Surveying and Mapping	71
9. Drilling and Blasting	73
10. Mine Support	79
11. Haulage	87
12. Hoisting	97
13. Mine Drainage	105
14. Mine Ventilation	109
15. Other Mine Services	115
16. Mineral Processing	119
17. Mine Management	121
Conversion Table of Units	127
Glossary of Mining Terms	129
Index	137
About the Author	143

1. Introduction

What is 'mining'?

What is the nature of the operations involved in the various fields of mining activity?

Well, of course, we could start with a definition:

> Mining is the process of extracting minerals of economic value from the earth's crust for the benefit of mankind.

Here we can assume that the earth's crust, the outer surface of the earth, including the oceans, lakes and rivers, extends to depths of 20 to 30 miles (say, 30 to 50 kilometres) or more.

But before we proceed further, let us see where mining fits into our economic system. What part does mining play in contributing to our national output of goods and services (the Gross National Product)?

The Structure of the Nation's Economy

As with most free-enterprise nations, the structural framework of our industrial economy embraces three different groups, viz.:

1. Primary Industry, which includes Agriculture (covering also forestry and fisheries), and Mining. These are the only two basic primary industries in that the goods they produce are derived from the earth's crust, including the oceans. In turn, the products of Agriculture and Mining become the raw materials for secondary industry. (But some agricultural products, such as fresh vegetables, are used directly by consumers).

2. Secondary Industry, in which the raw materials used as feedstock are the products of the two primary industries (either of domestic or foreign origin), and which are processed by various manufacturing techniques. As an example, agricultural products are typically fed to food processing plants, and to the paper-making, the building and construction industries, and other market outlets. Similarly, the mineral products of the mining industry are utilized in a number of vital ways by a wide range of processing industries. These cover power generation, steel production and fabrication, nonferrous metal fabrication, fertiliser production, and various processing techniques for the chemical, pharmaceutical, building, decorating, communication, cosmetic, defence and electronics industries. These techniques embrace the manufacture of all types of machine assemblies, such as farm implements, automobiles, appliances, aircraft, ships, and many other items of machinery and tools.

The above lists are by no means exhaustive. They are mentioned merely to give some sort of indication as to where Secondary (Manufacturing) Industry derives its raw material feedstocks: from the Agricultural and Mining Industries.

But, of course, no complex economy could operate with materials produced and stored at the factory door. In order to distribute these manufactured products among the consuming sectors of the public, we have a third component of the economy. It produces services.

3. Tertiary Industry involves:

(a) the sales and distribution (wholesale and retail) sectors of industry;
(b) the various transportation sectors, by land, sea and air;
(c) professional services, including health, law, education and religion;
(d) services of the various levels of government;
(e) maintenance services, such as the various repair trades, laundries, petrol stations, hairdressing and the like;
(f) the entertainment industry; and among others,
(g) financial, banking and insurance services.

In our complex industrial economy, it is of course obvious that none of the above three industry groups (Primary, Secondary and Tertiary) could be effective if operating alone. All three groups are interdependent in a highly developed society.

It is true that some early tribal communities were able to exist, as 'peasant economies', on agricultural output alone. But today, in most industrialized nations, even Agriculture depends upon Mining for fertilisers, chemical sprays, and machinery. To be sure, Mining also depends upon Agriculture for food, for many housing materials, and for clothing made from natural fibres (three of the prime necessities of life), as do the members of all industry groups, and humanity in general.

Nevertheless, although humanity may be able to exist (barely) on food, shelter and some clothing in a peasant economy, provided air and water are also available, the existence and maintenance of any industrialized society involves every member of that society in a high level of dependence upon mineral production (i.e. on mining activities).

It is interesting to note that at least one food item (salt) is recovered by mining techniques, as also are the raw materials for many medicines, for most plastics, and for all surgical instruments.

Therefore, it is very important for all members of our society to become actively interested in their mining industry.

Now we have two more questions:

What is a mineral?

We will deal with this question in a simplistic way because it must be assumed that some readers have no knowledge of general geology, and very little of basic chemistry.

Let us consider the question of distinguishing between (a) chemical elements, (b) minerals, and (c) rocks.

(a) Chemical Elements

Chemical elements are particles of matter existing as distinctive atoms. Each specific element has its own atomic weight and other characteristics such as specific gravity (density). All these elements are related to one another in a framework of characteristics known as Mendelèef's Periodic Table of the Elements. There are about 104 such elements that have been isolated and recognized in the laboratory. But most of them are found in the field in the combined form: combined with other elements.

Some elements occur in the natural state in greater quantities than others. For instance, 99 per cent of the earth's crust consists of only eight such elements, as set out below, with the chemical symbol of the element shown in parentheses:

Oxygen	(O)	47%
Silicon	(Si)	28%
Aluminium	(Al)	8%
Iron	(Fe)	5%
Calcium	(Ca)	3.6%
Sodium	(Na)	2.8%
Potassium	(K)	2.6%
Magnesium	(Mg)	2.1%

This means that all the other 96 elements are relatively scarce in that they represent only one per cent by weight of the earth's crust.

In fact, some of the more important metals (apart from aluminium and iron) that we must rely upon so greatly to sustain our standard of living, represent very minute proportions of the earth's crust, e.g. Copper (Cu), 0.0045%; Lead (Pb), 0.00015%; Gold (Au), 0.0000007%.

It will be seen that these and other important metals are dispersed very thinly in the earth's crust. At such low concentrations, we cannot afford to recover them (mine them) for the benefit of society, any more than we can afford to mine aluminium and iron at average concentrations of eight and five per cent respectively.

Well, how *do* we get over this problem? How is it that these metals are being mined? The answer to these questions is about to be unfolded here.

But now let us look at another aspect.

With few exceptions, these metals do not occur as elements. More usually, they are found combined or compounded with other elements, as we shall see; not the sort of chemical compounds that we find prepared artificially in the laboratory, but chemical compounds as found in the natural state in the earth's crust. These are called 'minerals', which we will deal with next.

But first we will refer to those exceptional cases of elements that occur 'native', i.e. in their natural state. There are very few. Some are:

Copper	(Cu)	Sulphur	(S)
Silver	(Ag)	Graphite	(C)
Gold	(Au)	Diamond	(C)
Platinum	(Pt)		

and the elemental gases of the atmosphere (O, N, etc.).

(b) Minerals

We may define minerals as "naturally occurring inorganic compounds with relatively precise chemical compositions and distinctive physical properties".

In a simple way, we may think of them as 'aggregates' of chemical elements, although this may not be a scientifically acceptable term. There are about 2,000 catalogued species.

Each particular mineral has its own mineralogical name which is quite distinct from the name of the artificially prepared chemical compound that we find in the laboratory. Although the chemical composition may be the same, the physical characteristics are usually quite different. For instance, lead sulphide, a chemical compound of lead and sulphur, with the chemical symbol of PbS, is generally available in the laboratory as an amorphous powder. It has quite different physical properties from 'galena' (PbS), the name of the naturally occurring mineral, often in crystallized or crystalline form.

One particular mineral deserves special mention, perhaps because of its widespread occurrence. It is 'water', a chemical compound of hydrogen and oxygen.

There are two exceptions to our general definition of a mineral as expressed above.

(1) There are some minerals which occur native, or uncombined, or alloyed with one another, such as native copper, native gold, etc. as mentioned above. These do not quite fit our definition.

(2) There are other naturally occurring materials that are not inorganic in origin and have no precise chemical composition and no definite physical characteristics. But for convenience we consider them as minerals. They are the naturally occurring 'organic' hydrocarbons, such as coal and petroleum.

Minerals fall into two categories for the purpose of our study, (a) common rock-forming minerals, and (b) 'useful minerals' (minerals of economic value). When sufficiently localized or concentrated by nature in what are known as 'mineral deposits', these latter are said to be high enough in grade to warrant extraction by mining.

(c) Rocks

Rocks are classified as *igneous, sedimentary,* or *metamorphic,* but these distinctions need not concern us here.

Actually, igneous rocks (such as granite and basalt) have emanated from the depths of the earth, and have become solidified from a molten state in the process. Sedimentary rocks are formed chiefly from the weathering products of older rocks by oxidation and weathering caused by atmospheric agencies such as the sun, wind, rain, ice and snow. The resulting fragments in time become deposited in the beds of lakes and the oceans as sediments, and have later consolidated and cemented under overlying pressure for millions of years. These consolidated beds (such as limestone) are sometimes found to contain the organic remains of plant and animal life (fossils). Metamorphic rocks are igneous or sedimentary rocks that have subsequently become altered (metamorphosed) by intense heat and pressure over long periods of time. Typical of these are gneiss, slate and schist.

Fig. 1 shows a typical chain of processes by which rocks of different types are formed, degraded, reformed and altered. Whatever the original disposition, rocks can become tilted, folded, and faulted by large scale movements in the earth's crust over many millions of years.

Whereas minerals may be considered as 'aggregates' of chemical elements, so rocks are really aggregates of minerals. They contain minerals in varying proportions, and therefore they have no definite chemical composition.

In general, rocks are of no particular economic importance, except in the case of those with desirable physical characteristics, such as construction stone, building stone, and monumental stone. Such rocks are mined for these purposes.

Two mono-mineralic types of rock should be mentioned. One is generally referred to as 'limestone'. It is basically a mineral (calcite, $CaCO_3$) in a crude impure form, but it occurs in such widespread

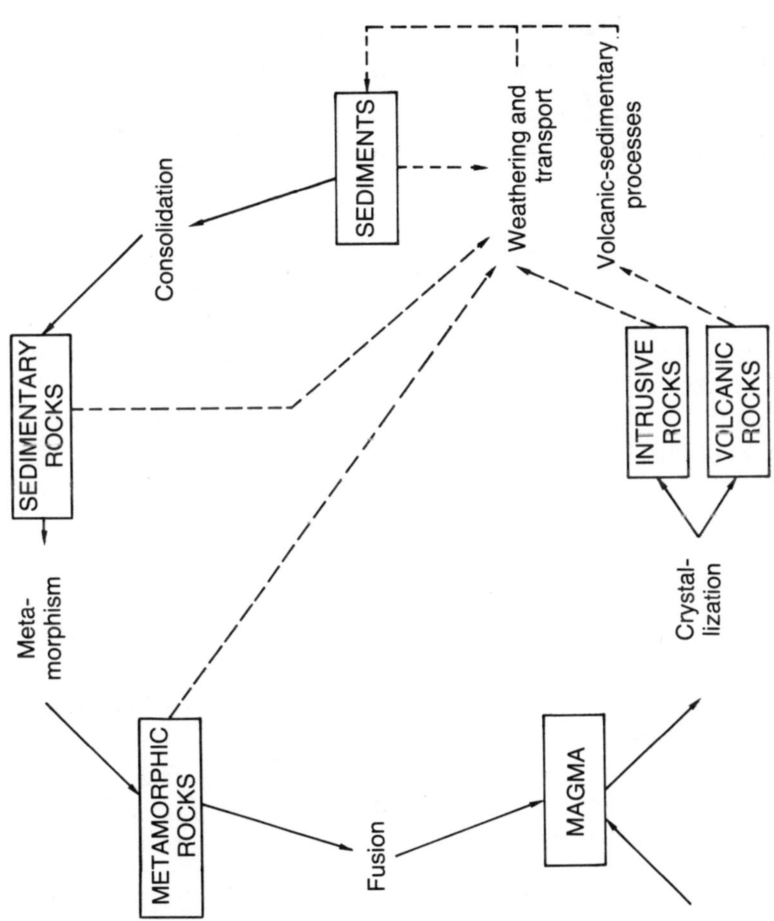

Fig. 1:
A Simplified Diagram of the Rock Cycle

deposits that it is generally regarded as a sedimentary rock. Limestone is mined as a very important raw material for cement manufacture, and for other purposes.

Another variety of calcite, occurring in large deposits, is 'marble', generally regarded as a rock, but usually pure enough to be rated as a mineral.

The other main mono-mineralic rock is known is 'sandstone'; it is really a large mass of consolidated crude quartz (silica) grains.

Coal is perhaps more accurately regarded as a carbonaceous *rock* as is oil shale. But in no way can we consider liquid petroleum (crude oil) as a rock, so we now accept these three main organic hydrocarbon materials as special types of minerals, as described above.

So now we have a reasonably clear idea of the composition of the earth's crust in terms of its separate classes of constituents. From now on we will omit serious consideration of the gases in the atmosphere, or the waters of the oceans, lakes and rivers, although these continue to play their part in the slow natural geological processes of weathering of rocks, which they convert into soil and other sediments. But they are not of major direct interest as economic materials to be mined under present conditions.

What is a Mineral of Economic Value?

The real economic targets of mining activity are about one hundred of the 2,000 or more catalogued mineral species, including the native minerals, the hydrocarbon 'minerals', and the few types of economic rocks mentioned above.

But in terms of our previous discussion, how can we consider mining these materials if they are so widely dispersed among the barren uneconomic rocks? This is the next important point to consider.

Actually, the dispersion of minerals throughout the earth's crust is not an even uniform phenomenon. In places we find useful minerals concentrated or localized by nature in various modes, as we shall see later. For instance, fractures in the rock mass may serve as channels for the deposition of mineral-bearing solutions from which the minerals are then precipitated to form what we call 'veins'. If the fractures are wide zones of weakness in the rock, we often find wider deposits called 'lodes' (see Fig. 2). There are other modes in which mineral deposits are found to occur.

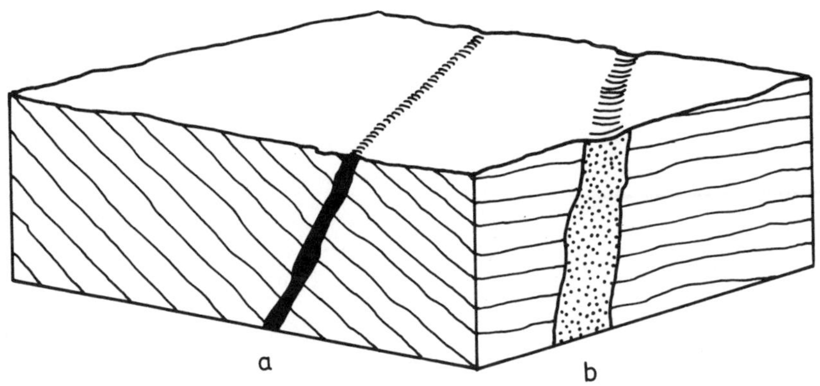

Fig. 2:
Typical Modes of Occurrence of Veins
(a) and Lodes (b) in the Crustal Host Rock

Under our private enterprise system, only those mineral deposits are likely to be mined that are sufficiently concentrated by nature (i.e. of high enough grade) to be extracted at a profit. And mineral deposits can be mined only where such deposits are found to exist, however inconvenient the location.

Recommended reading:

BLYTH, F.G.H. and de FREITAS, M.H. *A Geology for Engineers*, 6th edn. (London: Arnold), 1974, Ch. 1, 3, 4, 5, 6.

LEFOND, S.J., ed. *Industrial Minerals and Rocks,* 4th edn. (New York: AIME). 1975.

MASON, ANITA. *The World of Rocks and Minerals.* (New York: Larousse). 1976.

PETERS, W.C. *Exploration and Mining Geology*. (New York: Wiley). 1978.

READ, H.H. and WATSON, J. *Introduction to Geology*, 2nd edn. Reprinted by Macmillan, 1971.

2. Mineral Deposits

Classification of Economic Minerals

There is a broad range of 'useful' minerals that, after mining and processing, are customarily used as raw materials for the secondary or tertiary industries.

This range of economic minerals can be classified under the following headings. The list is by no means exhaustive.

(a) Minerals that yield metals

Precious metals: gold, silver, platinum.
Base metals: copper, lead, zinc, tin.
Steel industry metals: iron, nickel, chromium, manganese, molybdenum, tungsten, vanadium.
Light metals: aluminium, magnesium.
Electronic industry metals: cadmium, bismuth, germanium.
Radioactive metals: uranium, radium.

(b) Non-metallic minerals

Insulating materials: mica, asbestos.
Refractory materials: silica, alumina, zircon, graphite.
Abrasives and gems: corundum, emery, garnet, diamond, topaz, emerald, sapphire.
General industrial minerals: phosphate rock, limestone, rock salt, barite, borates, felspars, magnesite, gypsum, potash, trona, clays, sulphur.

(c) Fuel minerals

Solid fuels: anthracite, coal, lignite, oil shale.
Liquid fuels: petroleum oil, natural gas.

Classes of Mineral Occurrence

We may rate the relative possibility of economically mining deposits of useful minerals in three broad classes as set out below.

Class (1) Natural concentrations of economically useful minerals. They may be of relatively high or low grade, but nevertheless capable of being mined profitably. Some of these we know about; but there are obviously many other such deposits that are still undiscovered. This is the class that represents our present mining target.

Class (2) Mineral concentrations of sub-marginal grade, whether known or as yet undiscovered. At present costs and levels of technology and mineral market prices, this class is generally too low in grade to justify mining operations. In future years, as Class (1) occurrences become exhausted, this class may fit into our target.

Class (3) Dispersed minerals throughout the rocks of the earth's crust. These may be (a) common rock-forming minerals, such as quartz, felspars, and silicate minerals, or (b) potentially useful minerals. But these latter are too extremely widely dispersed and low in grade, and therefore of no economic value at prices ruling at present or in the foreseeable future.

Therefore it will be seen that the production of minerals of economic value by mining techniques must be confined to Class (1) mineral occurrences.

Our next approach is to study the various modes in which these localized mineral deposits occur. But before we go too far, we had better acquaint ourselves with the particular terminology used in the mining industry. Definitions of mining terms are listed in the *Glossary* at the end of the book.

Modes of Occurrence of Mineral Deposits

There are several accepted methods of classification of economic mineral deposits that have been localized or concentrated by nature. But *we* will classify them according to the mode or form of the deposit. Three modes of occurrence are recognized.

Veins are simply fissures, fractures or fault planes in a rock mass that have been filled by precipitation of minerals from solutions or by injection of magmatic mineral material from the depths of the earth.

Lodes, which are in zones of fractured rock, are similarly formed. Because these fissures and fractures generally occur radially, most veins and lodes have a near-vertical habit. They may be of several hundred feet (metres) in strike length, and the width is generally variable. Veins may bulge or pinch out over quite short distances in strike or depth. Lodes are usually much wider and of lower grade material (see Fig. 2).

Bedded deposits are typified by coal seams which are interbedded between the layers of stratified sedimentary rocks. Coal seams were formed from plant growth followed by a gradual sinking of the land below sea or lake level. The plant life then decayed and peat was formed. Sandy and muddy sediments settling through the water over periods of millions of years became highly compressed and gradually formed sedimentary rocks. The pressure from these overlying sedimentary rocks converted the peat into lignite and much later into coal. Seams of coal can be seen outcropping between beds of sedimentary rock in many of the highway cuts in the eastern United States. Most of the coal was formed from plants that grew about 250 million years ago. Most coal seams are in horizontal beds, but many have subsequently become tilted at various angles.

Replacement deposits also occur mainly in a more or less horizontal habit. They are metalliferous deposits that were formed by the dissolving away of sedimentary beds of limestone rock by mineral-bearing solutions which partly replaced the rock.

This phenomenon is also known as 'metasomatic replacement'. The mineralization introduced appears to occur in random sections of the beds replaced, but in a more or less tabular fashion. This is necessarily a simplistic explanation of the formation of replacement deposits.

In between coal and metalliferous mineral deposits we have a wide range of industrial mineral (generally non-metallic) deposits. Some of these deposits occur as veins, and others as stratified (bedded) deposits.

Bedded deposits may extend over large areas. Their lateral extent is usually much more widespread and continuous than metalliferous non-ferrous deposits. They contain minerals such as coal, iron ore, salt, potash, gypsum, and phosphate rock. Limestone, required for cement manufacture and for use as a smelter flux, is also mined from bedded deposits.

When the relatively soft industrial minerals are mined underground in bedded deposits, coal mine practice is generally followed. However, metalliferous (replacement) deposits in the horizontal mode are in hard rock. These need to be mined with explosives.

Grades of Ore

As will be seen from the definitions listed in the Glossary, the grade of an ore deposit is the equivalent average weight of metal contained in a ton (tonne) of the ore, even though the metal is in one or more of its mineral forms.

The ore grade is therefore of considerable economic significance because the metal recovered from every ton (tonne) of ore must yield enough revenue to cover the cost of mining, milling, and smelting that ton (tonne) of ore, plus overheads; otherwise no profit will ensue.

This means that attention must be paid to ensure that only ore of greater than the minimum economic payable grade is mined. This value is known as the 'cut-off grade': a grade that is marginal between a profit and a loss. Sub-marginal grade ores can be mined but only at a loss. In simple terms it can be said that the lower the unit working costs, the lower the cut-off grade, and therefore a certain amount of previously sub-marginal grade ore can now be profitably mined. This represents an increase in national wealth. On the other hand, additional government taxes add to costs, raise cut-off grades, and therefore reduce national wealth by restricting the tonnage of ore available for mining.

Table I shows that the various methods of mining typically involve different unit costs, ranging from square-set mining (the highest unit cost method) to surface mining (the lowest). Therefore it follows that the lowest possible cut-off grade will apply to surface mining (open cut work).

If however, the particular mineral deposit is situated well below the surface, it will need to be worked by one or other of the applicable underground mining methods, which are more expensive. In this case, the cut-off grade will need to be higher.

Nevertheless, because the various metals are priced at different dollar values day by day on the world markets, and for other reasons, it also follows that ores containing single metals will have widely different cut-off values.

Typical cut-off grades (that yield a minimal profit) for some of the metals are shown below. They are expressed as for open cut mining, and for metal prices ruling generally in 1976.

Antimony	4.5%
Copper	0.5%
Chromium	35%
Iron	28%
Lead	2%
Manganese	28%
Nickel	0.5%
Tin	0.45%
Zinc	3%
Gold	0.15 oz/ton (5 grams/tonne)
Platinum	0.12 oz/ton (4 grams/tonne)
Silver	1 oz/ton (30 grams/tonne)

Table I — Comparative Direct Costs of Mining (1976)

Method	Average cost per tonne ($)
Square setting	20
Cut-and-fill	14
Shrinkage	12
Sublevel longhole	9
Open longhole	5
Block caving	3
Open cut	1

The foregoing discussion applies to what are known as 'straight' ores: ores containing minerals of a single metal. It is interesting to note that the cut-off grade of straight gold ores is about one-sixth of a troy ounce per ton. This is about the size of a 5-cent coin, recovered from 12 cubic feet of ore (about one-third of a cubic metre) weighing about one ton (tonne).

However, many metalliferous minerals occur in close association with one another, and therefore several marketable products may be recovered at little extra cost from the same ton (tonne) of ore. For instance, we find that lead and zinc sulphide minerals frequently occur in the same orebody, and some silver is also present. Other complex ores may also contain such metals (in mineral form) as copper, antimony, molybdenum, and pyrite.

This means that a ton (tonne) of this ore will yield greater revenue and the mean cut-off grade can therefore be reduced.

Because orebodies seldom contain the useful minerals in an evenly distributed pattern, it will be understood that some sections carry richer values than others. This feature dictates the need to mine the respective tonnages of richer and poorer ore that will yield a blend of the regular on-going cut-off grade desired. Inevitably, some submarginal grade ore will be left unmined. Whether or not this ore can be mined in later years depends upon the future level of cut-off grades that may then be applicable. In many cases, it is irrevocably lost.

There are many other interesting features. It so happens that most mineral deposits containing high-grade ore are small in dimensions. These need to be worked on a small scale by relatively costly manual methods.

On the other hand, many massive deposits (such as wide lodes) of low-grade ores have been found in recent years. These can now be worked profitably by mass production methods, mostly on the surface but also underground, because more space is available for mechanized equipment.

Recommended reading:

Peters, W.C. *Exploration and Mining Geology.* (New York: Wiley). 1978. Ch. 7 and pp. 68—70, 126—7.

Thomas, L.J. *An Introduction to Mining*, Rev. edn. (Sydney: Methuen of Australia). 1978. Ch. 1.

3. Three Classes of Economic Minerals

Fig. 3 shows the main distinctions between the mining and processing of metalliferous minerals, industrial minerals, and fuel minerals (typically coal), up to the point where they reach the market.

We see that **metalliferous minerals** occur in 'ore deposits': in veins or lodes in a near-vertical habit, or as bedded deposits in horizontal formations. In each case, the ore and the enclosing rock are very hard, and can be mined only with the use of explosives.

In a typical orebody, the valuable minerals occur in a groundmass or matrix of other worthless minerals called 'gangue'. In other words, the gangue may represent 95 per cent or more of the total amount of ore mined. The valuable minerals are usually disseminated in tiny specks, blebs or veinlets throughout the gangue.

The main problem is to beneficiate the ore (or to concentrate the valuable minerals) by separating and discarding the gangue as early as possible in a *mill* (concentrator, or treatment plant) set up on the surface. This can be done after first crushing and grinding into very fine particles the lumps of ore delivered from the mine. This fine grinding is necessary to produce separate particles of valuable mineral and gangue material. The actual physical separation of the bulk products (valuable mineral and gangue) is then achieved by one or other of many alternative processing methods, generally referred to as 'concentration', because the separation and discarding of the gangue minerals as tailings leaves behind a 'concentrate': a much smaller bulk of material containing practically all the valuable mineral. This separation process is not perfect in practice because a small fraction of the valuable mineral is lost in the tailing. But the concentrate is still in the form of a mineral which has only a limited sale value in the market. The next step is to extract the metal from the mineral concentrate.

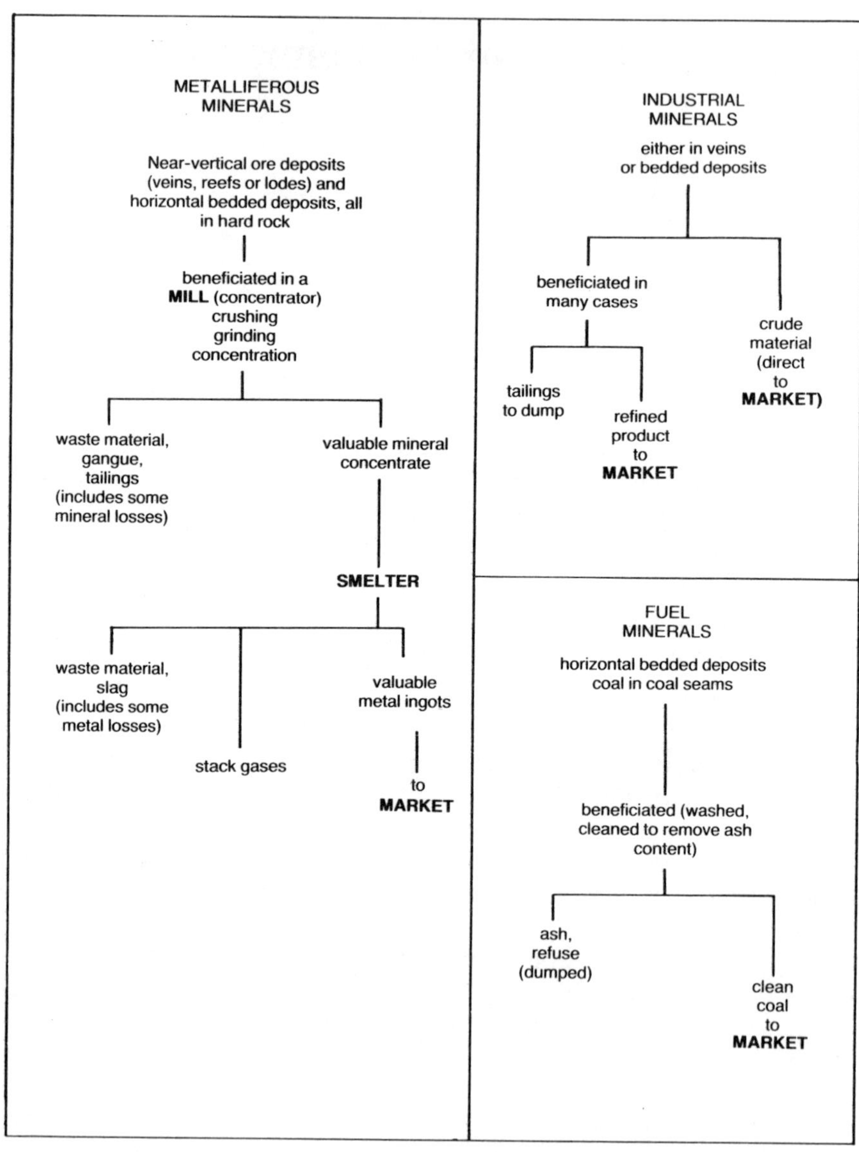

Fig. 3:
Three Aspects of Mining Activity

The processing unit in which metal extraction is performed is typically called a *smelter,* established on the mine site, or perhaps at some industrial centre or port, where fuel and electric power are available at a reasonable cost.

Many different smelting processes are employed for the reduction of metalliferous ores and/or concentrates to produce metal ingots. They are specific for each metal.

In an iron or lead smelter, the concentrate, along with fluxing agents and coke fuel, are fed to a blast furnace. In time, the mineral concentrate is reduced to a molten metal collecting at the bottom of the hearth. Some of the impurities collect and float above this as a fluid 'slag', and others pass along the flues and up the stack as gases (typically sulphur dioxide) and fine dust, which are now being recovered or purified before being released to the atmosphere. At intervals, the molten slag is tapped and run off into a suitable receptacle, granulated into small pieces with a stream of water, and stored or dumped for sale as rubble or roadmetal. Similarly, the molten metal bath is tapped at intervals and run into the specially shaped ingot moulds of a casting machine. On cooling, the solid ingots of metal (in this case, pig iron or lead) are shipped to market. Again, not all the metal contents of the concentrate are recovered.

There is a wide range of **industrial minerals** mined for a myriad of industrial end uses. They may occur in near-vertical vein deposits, or in bedded deposits, or in other modes. The run-of-mine material delivered to the surface may be marketed direct, where the end use calls for a crude cheap product (as for barite, used as a drilling mud in oil well exploration work); or it is beneficiated to remove the impurities where a high technical grade product is required. The beneficiation plant and methods are based upon the particular chemical process reactions required to remove the impurities before the cleaned product is delivered to the market. The impurities are seldom more than 20 per cent of the mined material.

With **fuel minerals** the proportion of worthless material, generally called 'ash', is usually no more than 25 per cent by weight. This represents shale and earthy materials (sometimes pyrite) which are unavoidably mined along with the coal.

The valuable constituents of coal are the combustible carbon and volatile gases that together represent the heating potential or 'calorific value'of the particular coal. The incombustible ash obviously makes no such contribution. If the coal needs to be transported great distances to market, then it will be surely advantageous first to remove the ash in order to economize on freight charges.

Ash is therefore removed by a beneficiation process, in which the run-of-mine coal is beneficiated (cleaned or washed). This is performed in a cleaning plant (or coal washery) erected on the surface. The ash is dumped as 'refuse' or 'slag', and the cleaned coal is loaded into railroad cars or river barges for despatch to the market.

Because coal has no definite chemical composition like regular minerals of inorganic origin, it cannot readily be evaluated in terms of 'grade'. It is usual to evaluate (or to 'rank') a coal in terms of its fixed carbon content; or by its calorific value: the number of heat units produced by unit weight of that particular class of coal, when free of extraneous matter such as moisture, mineral matter and ash.

The various ranks of coal are tabulated in Table II below, on a moisture, mineral matter and ash free basis.

Table II. Classification of Various Coals

Type of coal	Description	Fixed carbon content (%)	Calorific value BTU/lb	Cal/g	Joules/g
Anthracite	Hard coal	90+	15,000	8,300	35,000
Bituminous	Soft black coal	85—90	13,000	7,200	30,000
Sub-bituminous	Grey coal	80+	10,000	5,500	24,000
Lignite	Brown coal	75+	7,000	4,000	16,300
Peat	—	—	3,000	—	—

Recommended reading:

PETERS, W.C. *Exploration and Mining Geology*. (New York: Wiley). 1978.

THOMAS, L.J. *An Introduction to Mining,* Rev. edn. (Sydney: Methuen). 1978.

4. Mineral Exploration

One of the most fascinating aspects of mining is the search for and discovery of mineral deposits (naturally concentrated or localized occurrences of useful minerals). This is termed **prospecting**.

In earlier times, prospecting was carried out by individual prospectors, using rule-of-thumb methods and a certain amount of intuitive knowledge, combined with a minimum range of tools and camp gear, in mineralized terrain. The general aim was to locate the outcrop of a vein or lode, to expose it across its width, along its strike and for a limited depth (by sinking pits at intervals). Many such outcrops were partly buried and had to be exposed by cutting 'costeans' laterally across the vein at intervals. By sampling the exposed faces and having assays (analyses) made, the average grade of the deposit was roughly deduced.

Except perhaps in regions where arctic ice, desert sands or tropical rain forests exist, most outcropping mineral veins have already been discovered. For this reason, the role of the individual prospector is rapidly diminishing.

When prospecting for vein gold, a prospector typically washed the gravels in stream beds in a gold pan, a manual 'tool' for separating heavy mineral particles from the lighter stream gravels and silt. Where evidences of gold, collected in the pan, persisted, he would work his way up the stream bed, progressively panning along the way, in the hope of finding its source (the 'mother lode'), whose outcrop was often concealed by the weathering products of the hillside rocks.

The above are regarded as 'conventional' prospecting methods. In recent years, more sophisticated methods have been evolved, using instrumental procedures based upon the geophysical or geochemical sciences. They are applied by the larger organizations in mineralized areas that have been geologically mapped by aerial or surface methods. Airborne exploration is particularly adapted to rugged or difficult terrain. When outcrops are concealed, the exploration engineer then needs to apply geophysical procedures in support of surface mapping.

Geophysics is the application of physics to the study of the earth. Orebodies frequently differ in many of their physical properties from the surrounding rocks in which they are enclosed. The detection of some of these differences indicates the presence of an 'anomaly', which in turn may or may not signify the presence of a buried mineral deposit. These anomalies therefore provide targets for localizing search activities by exploration drilling. A survey aircraft may carry many different instruments to record variations in a number of physical properties, such as magnetic susceptibility, density, electric conductivity, etc. The various geophysical methods may exploit differences in magnetic, electro-magnetic, radio-active, gravity, chemical or seismic properties.

Geochemical methods involve analyses of samples of surface soils, lake water, and vegetation to determine values of elements having a background value higher than normal. In this way, an area of low grade mineralization may lead to the discovery of a significant orebody.

Aerial photography has been used to indicate large scale structural formations that may host mineral occurrences. More recently, satellite photography and other methods are used to indicate mineralized areas by colour infra-red photography, especially where orebodies are oxidizing significantly.

Radioactive mineral occurrences are located by the use of radiation counters, both airborne and on the ground surface. Typical of these are the Geiger-Müller Counter and the Scintillometer.

For bedded occurrences, unless outcrops are visible as in hilly terrain, deposits (mainly of coal and industrial minerals) are often located in the first instances by vertical holes drilled for water.

Any anomalies plotted from the results of geophysical prospecting are tested for the presence of useful minerals by drilling. There are several drilling techniques in use. A common one is the diamond drill.

Diamond drilling is the method used to penetrate rock at depth and to sample the downward extension of a mineralized vein or lode by recovering a core of the rock or vein material so penetrated.

A diamond drill unit consists of a drive motor, rope hoist and mast, water circulating pump and drive chuck. The drive chuck rotates a string of screwed hollow rods at the bottom of which is a core barrel and diamond-studded bit. Rotation of the rods causes the bit to grind an annular ring of rock at the bottom of the hole into small fragments, allowing a central 'core' to rise into the core barrel as drilling proceeds. The ground rock sludge flows to the surface, in the space outside the rods, under pressure from circulating water pumped down inside the hollow rods. At intervals, the core can be recovered from the core barrel and examined. If mineralized, the core can be assayed for its mineral content, as also can the sludge.

The second phase of mineral exploration activity is concerned with the work of **evaluation** of a mineral deposit found by prospecting (conventional or instrumental).

Evaluation relates to the determination of the profit potential of mining such a deposit in terms of the prospective quantity (tonnage) of recoverable mineral matter and the average grade of the contained mineral or minerals.

Tonnage is determined by a programme of diamond drilling to find the limits of the deposit in terms of its length, average width and depth. The total volume thereby derived, combined with the average density, yields the tonnage figure. By using the mineralized drill cores as samples for assay, the average grade can also be determined; but these samples may also be supplemented by face samples from shallow shafts, crosscuts and drifts.

The compilation of tonnages and grades over different sections of the mineral deposit (or orebody) is totalled and referred to as the 'calculation of ore reserves'. Ore reserves can be classified as:

(1) **Proved ore,** where the tonnage and grade of ore have been carefully outlined and assessed in three dimensions by excavation or drilling, but also include minor extensions where limiting geological factors are definitely known. These reserves are also described by some authorities as 'measured ore'.

(2) **Probable ore** (or 'indicated ore') applies to those parts of an orebody adjacent to proved ore where the conditions cannot be so precisely defined, but indicate that ore will probably be found.

(3) **Possible ore** (also known as 'inferred ore'), where insufficient quantitative estimates are available, and limited to an assumed

continuity based upon a broad knowledge of the geological character of the deposit. The reliability of this category is too low for inclusion in reported ore reserves.

Proved ore (and sometimes probable ore) gives a figure for the contained salable mineral or metal at the ruling market price. It enables the organization to plan the scale of operations and other features, based upon what is known as a 'feasibility study'. The feasibility report is used as a basis for decision making, as to whether to abandon the deposit, or to raise capital funds to develop the property and to mine the deposit.

Recommended reading:

DAVID, M. *Geostatistical Ore Reserve Estimation*. (New York: Elsevier). 1977.

PETERS, W.C. *Exploration and Mining Geology*. (New York: Wiley). 1978. pp. 475—488.

THOMAS, L.J. *An Introduction to Mining*. Rev. edn. (Sydney: Methuen). 1978. Ch. 3.

5. Sampling and Assaying

Few orebodies contain useful minerals regularly and uniformly dispersed throughout the mass. But most contain a mixture of minerals that vary in different parts of the deposit. In these especially it is paramount to take representative samples at close intervals in order to obtain an objective evaluation of the total metal content of the deposit.

Probably the best samples are obtained from diamond drill cores, especially where there is a high recovery of core along the intersection of the vein. This core from the vein is then split longitudinally and selected sections of one half are sent for assay (analysis). The length of each section of core, in inches or centimetres, is then related to its true width, and the sectional assay values weighted accordingly.

Otherwise, the exposed faces of an orebody are sampled from wall to wall at intervals along each face. This is performed by groove sampling (channel sampling), using a hammer and moil. A groove about three inches (8 cm) wide and two inches (5 cm) deep is cut across the vein from one wall to the other, and the moilings (chips, fragments and dust) are caught on a plastic sheet, section by section, and transferred carefully to a plastic sample bag, together with an identifying numbered tag torn from the sample book. Details of the position and true width of the sample are entered on the heel of the numbered pages of the sample book. Care must be taken to avoid taking a disproportionate amount of soft material where the ore face is banded with alternate hard and soft sections. This would introduce bias to the sample.

Sample bags are despatched to the Assay Office where the samples are crushed, reduced in size, prepared and assayed (whether by the dry or wet method). For routine mine operations, the assayer issues a

Daily Assay Report showing the assay values against the number of each individual sample. From these, Daily Assay Returns, showing also details of each sample, are prepared by the Geological Office for management and record purposes. Such assays are plotted on mine assay maps.

Once the face sample is taken, great care must be exercised in guarding it to prevent the introduction of extraneous material by interested persons with nefarious motives. This practice is known as 'salting the sample'. The objective is to give unduly high or low results that could benefit certain persons with a pecuniary interest.

The conventional method of assaying for gold and silver is by fire assaying; for base metals, by wet chemical analysis, as also for industrial minerals. Solid fuel minerals are evaluated by their calorific values. Modern instrumental methods offering faster, more accurate and more comprehensive results are now being used extensively.

Recommended reading:

McKinstry, H.E. *Mining Geology*. (Englewood Cliffs: Prentice-Hall). 1948. Ch. 2, 3.

Peters, W.C. *Exploration and Mining Geology*. (New York: Wiley). 1978. Ch. 16.

6. Four Classes of Mining Operations

A mine is an excavation made in the earth for the purpose of extracting useful minerals. Such a mine can be established on the surface or underground (see Fig. 4).

But an underground mine is not, as the man in the street believes, merely a hole in the ground. If it were, operations would soon have to cease for reasons of safety and economy.

On the contrary, to ensure that the mine will produce safely and economically throughout its working life, it must be *developed* systematically. This means the deposit must be sectionalized into local areas that can be conveniently and safely exploited without prejudice to the stability of neighbouring areas, and so that the ancillary processes of haulage, drainage and ventilation can be planned effectively in a logical layout.

Mine Development

The system of development depends upon the configuration of the deposit and on the planned scale of operations. It is different for a vein or lode than it is for a horizontal type bedded deposit.

Surface deposits are developed for long term objectives, based upon the ore reserves and the configuration of the orebody. The development of benches depends upon the grade of the ore and the geological structure, as shown on exploration maps, subject to various constraints that may apply, such as the physical nature of the ore and overburden, cost factors, and legal requirements. A specific development plan is therefore required for each individual ore deposit.

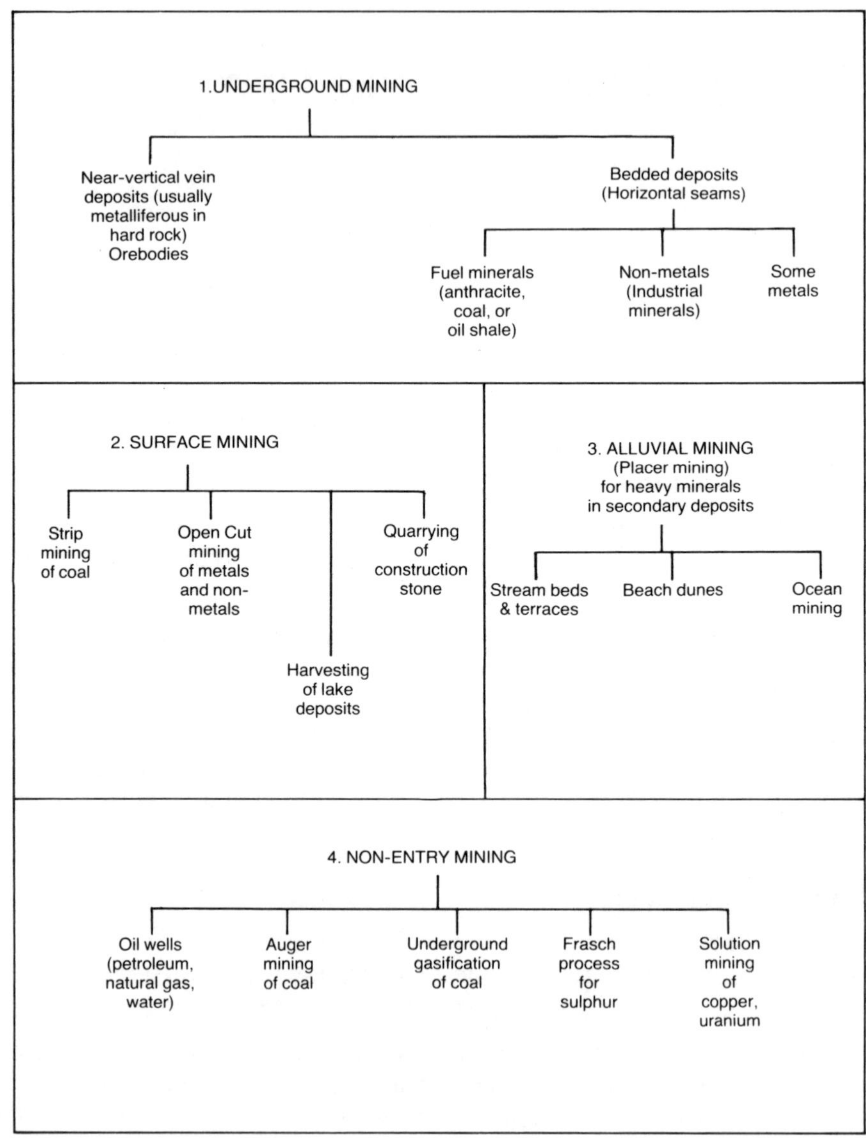

Fig. 4:
Four Classes of Mining Operations

The main access opening to any underground mine, whether a vein type or bedded deposit, is a **shaft**, where the surface terrain is comparatively level. In mountainous country, access to the deposit may be achieved more effectively by an 'adit' opening instead of a vertical shaft; or by a 'slope', or inclined shaft.

A vertical shaft may be rectangular, square, elliptical or circular in cross-section. With rectangular shafts, the rock walls are supported by heavy framed blocked timber sets (see Fig. 5), spaced about six feet (2m) apart. The timber supports the shaft walls and also provides a method of dividing the shaft area into compartments for various functional purposes. Similar timbering details are provided for square section shafts. For veins dipping at 60 degrees or less, some mines use inclined shafts. These are generally rectangular, with an arched roof, and for best effects are sunk in the footwall and generally parallel to the vein, but at a constant inclination. Rail-mounted skips are used, hoisted by steel wire ropes.

Elliptical shafts were developed from rectangular shaft layouts so that the walls could be supported by a monolithic concrete lining. The shaft is divided by buntons grouted into thc walls.

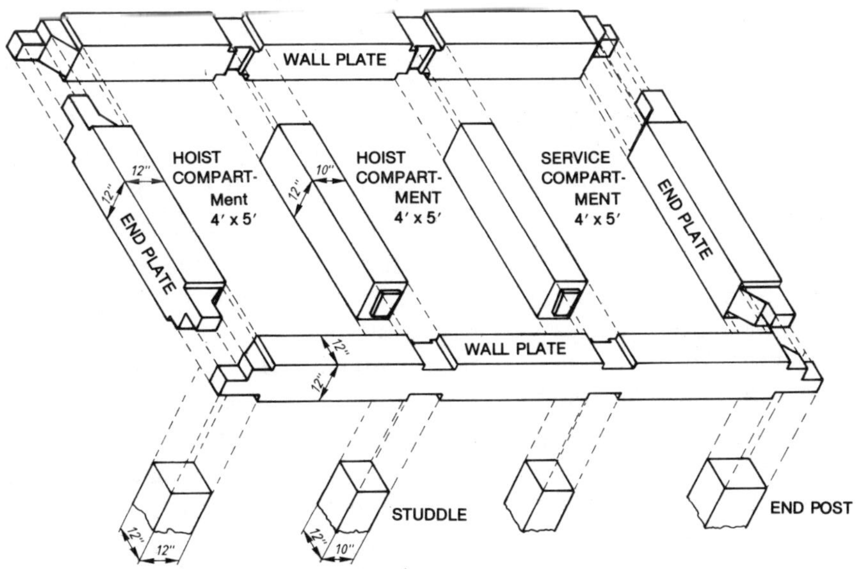

Fig. 5:
Timbered Shaft Set for Rectangular Shafts

Modern shaft practice now employs circular shafts which offer the greatest resistance to lateral rock pressure. Such shafts are lined with monolithic concrete and divided with steel bunton sets, at 10 to 16 ft (3 to 5 m) intervals, to which steel guides are connected for the cage or skip conveyances. Such shafts are now being sunk from 16 to 30 ft (5 to 9 m) in diameter, to depths of 7,000 ft (say 2,100 metres).

Since all fresh ventilating air should pass down one shaft and be exhausted up another, the resistance to airflow becomes an important matter. Timbered rectangular shafts offer very high resistance, but in concrete-lined circular shafts the resistance is minimal. Maintenance costs are also much lower in circular shafts.

Drifts and **crosscuts** have a 'tunnel-like' cross-section, with an arched roof, excavated in the rock and supported by timber sets wherever necessary. They vary in size from 5 x 7 ft (1.5 x 2 m) to about 10 x 14 ft (3 x 4.25 m), depending upon their particular purpose. Such an opening usually carries rail tracks, a floor drain, pipes and cables, and perhaps a ventilating duct. Where **adits** are used as main access openings (instead of a shaft) in hillside deposits, they are of similar construction. Drifts, crosscuts and adits are driven on a slight upgrade to assist water drainage.

As may be inferred from the Glossary, a **level** represents a system of horizontal connected openings (drifts and crosscuts) on one particular horizon, connected by raises or winzes to other levels, above or below. Levels are driven from the shafts at designated vertical intervals (100 to 300 ft, or say 30 to 100 m), depending upon a number of parameters, too complex to describe here. Each level provides a base for ore extraction, and for haulage of ore to the shaft for hoisting to the surface.

A **winze** is a vertical or inclined (down the dip of the vein) opening sunk from one level to another, or, for exploration purposes, below the lowest level. Winzes are generally about 5 x 7 ft (1.5 x 2 m) in cross-section. They are usually equipped with a small compressed air hoist, and with ladders, cables and pipes.

In apposition to winzes, **raises** are driven upwards from one level to another, generally along the dip of the vein. Conventional drill-blast methods call for some method of timbering with frame sets; or by using Alimak equipment (Fig. 6); or a cage method. More recently, raises are being bored mechanically by a pilot hole and reverse reamer equipment, from 4 to 8 ft (1.25 to 2 m) in diameter. When completed, raises serve the same general purpose as winzes.

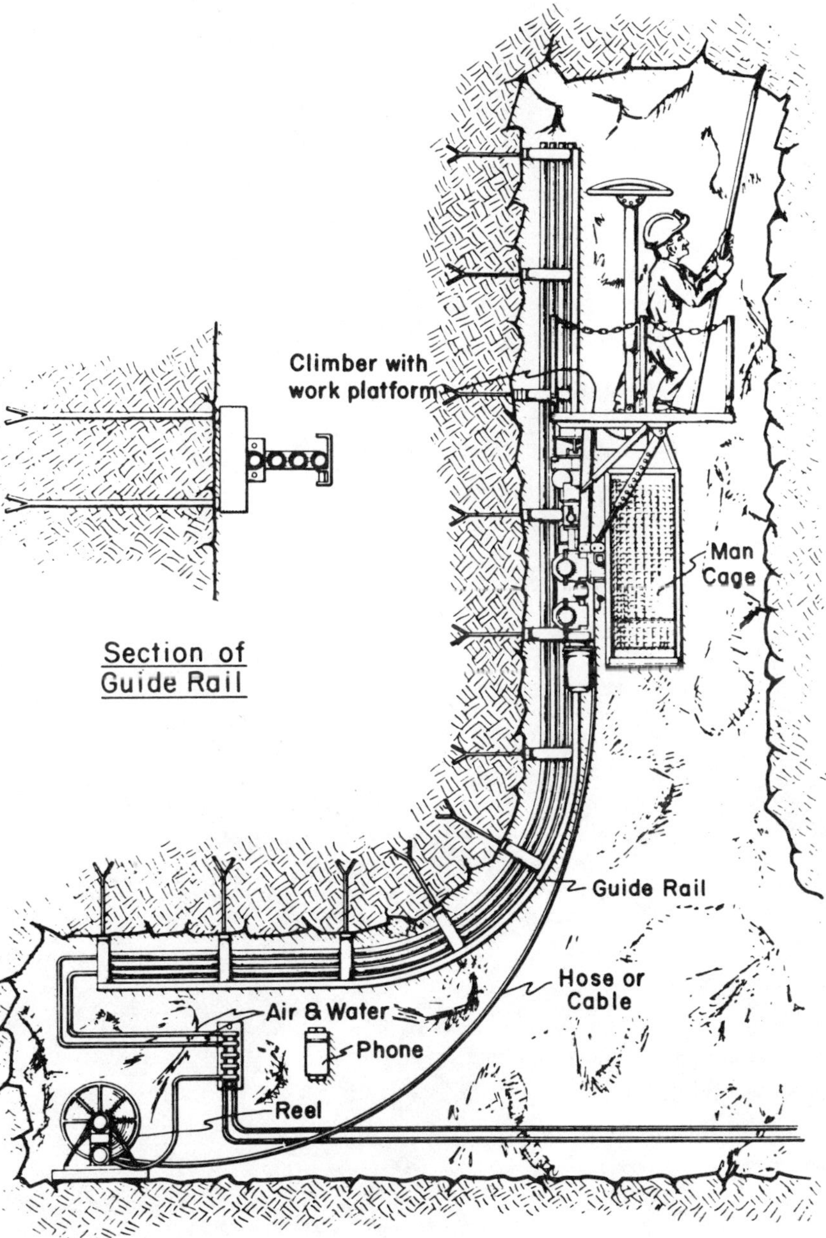

Fig. 6a:
Alimak Unit

Figs. 6 b, c:
Alimak Unit

Underground Mining

1. Near-Vertical Vein Deposits

The shaft is usually sunk in the footwall rock, in order to preserve the stability of the mine throughout its working life (see Fig. 7).

Such a shaft is divided into compartments through which particular operations are confined. For instance, a pair of compartments may be set aside as travelling ways through which skips (carrying ore) can be hoisted to the surface. Or alternatively, cages can replace skips for part of the working shift for hoisting or lowering workmen, material and supplies. These cages or skips are hoisted by means of stranded steel wire ropes powered by hoisting engines at the surface, with a headframe erected over the collar of the shaft.

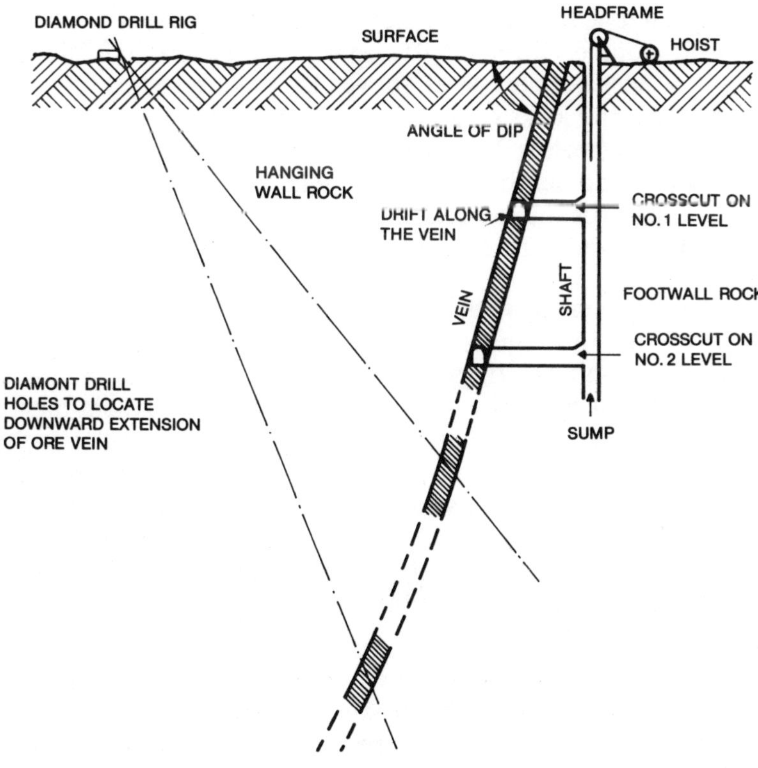

Fig. 7:
Cross-section of Hypothetical Ore Vein

Another compartment is designed to be equipped with ladders set on platforms or stages at invervals, and may also carry pipes and electric cables. All compartments provide an airway for the passage of fresh ventilating air into the mine. But this means another shaft is normally necessary to exhaust foul air from the mine. A large fan is customarily mounted over the collar of this second (ventilation) shaft to provide a sufficient flow of ventilating air through the mine.

For expository purposes, we are going to assume a hypothetical vein of ore 1,200 feet (about 400 m) long, with a regular average width of six feet (2 m), and dipping at about 75 degrees.

For such a mine, a crosscut is driven from the main shaft to and through the vein at vertical intervals of about 100 feet (30 m). Drifts are then driven along the vein throughout its strike length, which in our hypothetical case is of the order of 1,200 feet (about 400 m).

The workings on each of these separate horizons are called 'levels', and designated as such in numerical order. At intervals of about 200 feet (70 m) along each level drift, 'raises' are excavated in the ore of the vein to meet the next level above. By this procedure, the vein is sectionalized into blocks of ore about 200 ft (70 m) long and about 100 feet (30 m) high (see Fig. 8). These blocks of ore are therefore open or exposed at the top and bottom and at each end. Each such ore block provides a base for the extraction of ore by one or other of a number of classical methods of 'stoping'.

The development of such an underground metalliferous deposit is planned to suit the requirements of the particular stoping method selected. The stoping method should provide the most economic recovery of the useful mineral, consistent with safety. The main factors to be considered in selecting this method are (1) the physical characteristics of the deposit in terms of the vein width, the angle of dip, the relative regularity of the deposit and the strength of the ore, (2) the nature of the enclosing rocks, (3) the preservation of the natural environment, and (4) the total costs involved. Naturally, the operation must still yield a profit after subtracting the costs from the revenue derived, in any given period.

The hypothetical ore block shown in Fig. 8 cannot be removed *in toto* to the surface. It must be broken into small chunks. The ore is very hard and therefore it can be broken only by blasting with explosives.

In order to place the explosives within the mass of ore to break it,

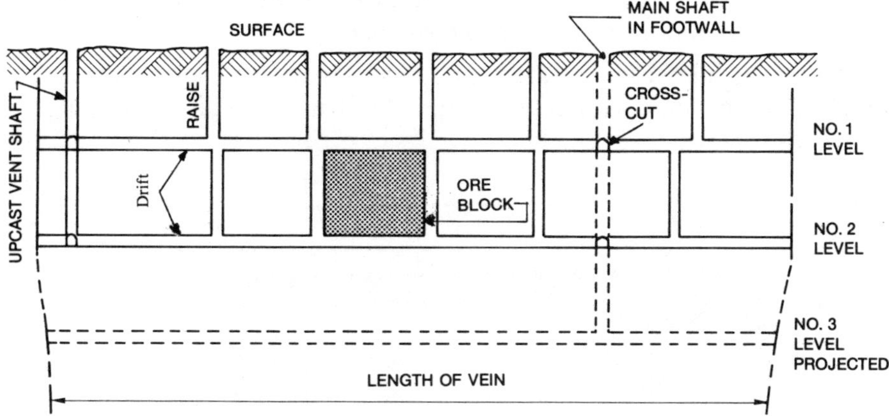

Fig. 8:
Longitudinal Section of Vein Shown in Fig. 7,
with Typical Development Openings on Two Levels

holes are drilled into the ore with pneumatic rockdrills. These blastholes are charged with explosives and fired at the end of each shift. The resulting dust, heat and gases are then expelled from the mine by the ventilation system while the miners are changing shift.

The broken ore chunks gravitate through chutes (see Fig. 9) controlled by chute gates to fill ore cars that are hauled in trains along

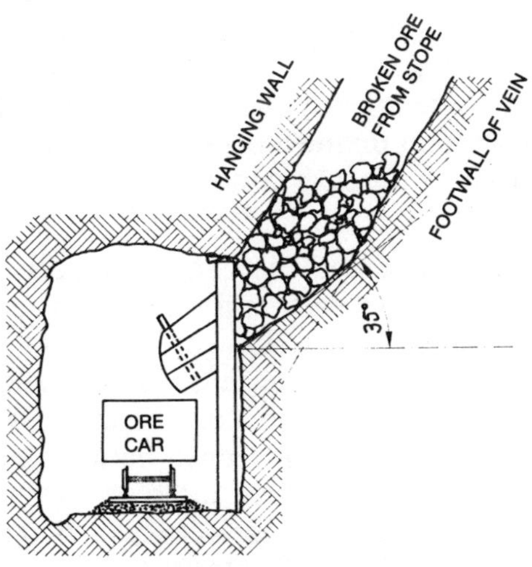

Fig. 9:
Simple Wooden Chute Gate

each level to the shaft. Here the ore is transferred into skips and hoisted to the main ore-bins on the surface.

One of the classical methods for steep narrow veins where the ore and wall rocks are reasonably strong is known as 'shrinkage stoping' (see Fig. 10).

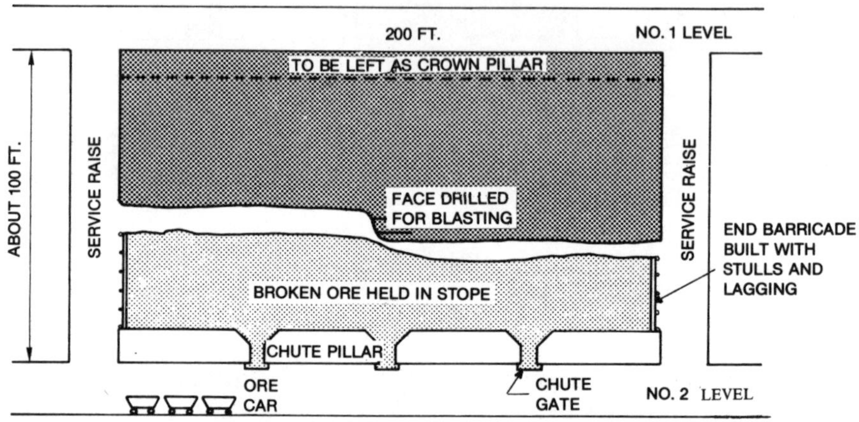

Fig. 10:
Method of Mining Ore Block by Shrinkage Stoping

Here the ore block is prepared by excavating a series of chute raises for a distance of about 12 feet (3.5 m) above the lower level. These are connected at the top by a sub-drift, and the entrance to each chute is funnelled out to provide an easy passage for the broken ore to pass through the chute without undue blockage. Chute gates are erected at the bottoms of the chute raises to control the flow of broken ore into ore cars (see Fig. 9).

Stoping is now ready to commence. In our hypothetical case, holes are drilled horizontally from each end raise and a quantity of ore is blasted down, some of which will flow directly into the nearest chute. To prevent the ore from spilling out into the main end raises, a barricade is provided at each end of the stope. This barricade consists of round timber 'stulls' (see Fig. 11) wedged between the hanging- and foot-walls at about five feet (1.5 m) intervals. Behind these stulls, hardwood planks are spiked in a near-vertical arrangement. As more ore is mined and the 'back' of the stope proceeds slowly upwards, more stulls and planks are placed, extending the barricade, but still

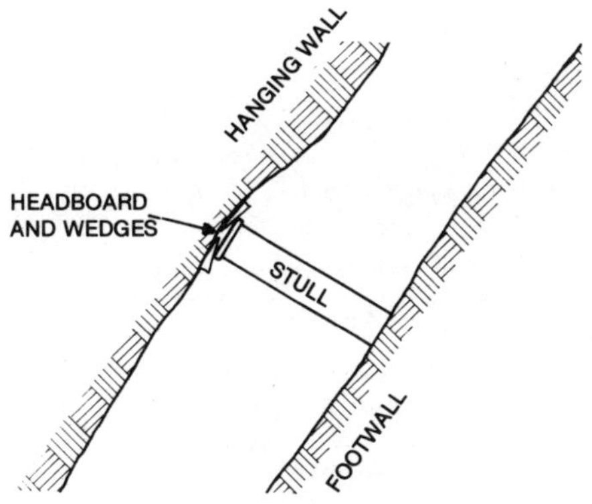

Fig. 11:
Round Timber Stull

leaving enough room for workmen with their tools to enter the stope, and for ventilating air to pass through.

In this way, slices of ore perhaps eight feet (2.5 m) high are stripped off the back of the stope. Sufficient broken ore is left in the stope to provide a working platform for the miners and to support the walls of the excavation. But about one-third of the broken ore is progressively drawn off through the chutes in a controlled manner to allow for the swell factor.

When the back reaches a point about ten feet (3 m) below the level above, the stope is completed, leaving a 'crown pillar' above to protect the upper level. At this point, all the remaining ore can now be drawn out through the chutes, and the resulting stope cavity can, if necessary, be filled with waste material to prevent collapse of the walls.

At a later stage of the life of the mine, the crown pillar and the chute pillars can usually be mined, by a special programme which we will not consider here.

The above is a simplistic description of a standard shrinkage stope. All sorts of variations have been adopted in different mining districts.

Another classic method of mining is known as 'cut-and-fill stoping', as shown in Fig. 12.

Here the ore block may be prepared in the same way, but the chute raises are not funnelled out at the top.

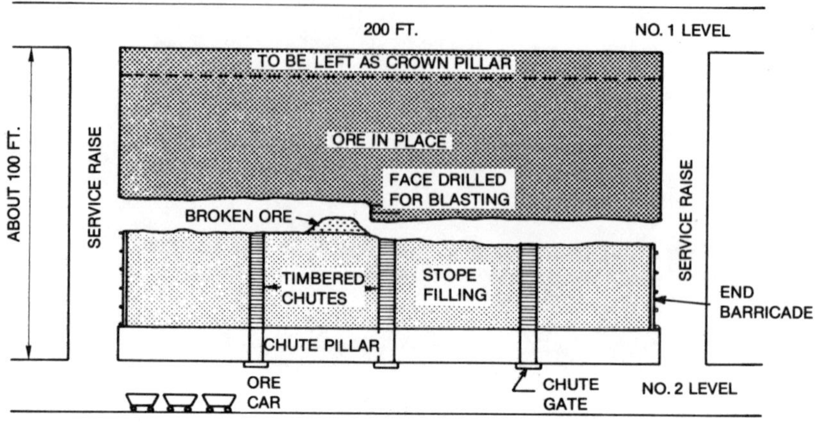

Fig. 12:
Method of Mining Ore Block by Cut-and-Fill Stoping

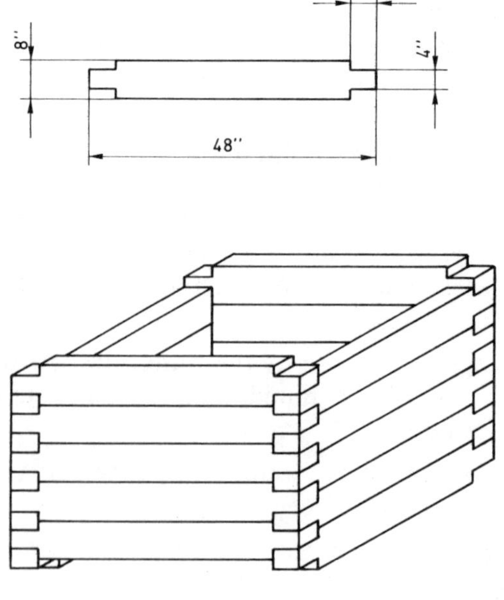

Fig. 13:
Chute Timbering

After every eight feet (2.5 m) slice of ore has been stripped from the back, a series of specially cut sets of standard timbers (see Fig. 13) are built up above each chute to within about 8 feet (2.5 m) of the back. Waste filling material is now placed in the stope between adjacent timbered chutes and between the end chutes and the barricades. Several alternative filling materials and methods may be used, but we will not enlarge on them here.

The top of each layer of filling now provides the working platform and the broken ore is loaded or scraped into the chutes.

As the stope proceeds upwards, timbering and filling proceed on a cyclical basis. When the crown pillar is reached, the stope is completed and abandoned. Various alternative types of cut-and-fill stoping are found in different mining districts.

There are several other standard methods of stoping narrow veins, but the above is sufficient to give an insight into the general stoping procedures.

Of course, for wide lodes and other configurations of orebodies, a number of other stoping methods are available, the choice depending mainly upon which one more closely applies to the physical characteristics of the ore and walls. In many of these cases, there is sufficient room to deploy mobile mechanized drilling and loading equipment under mass-production conditions.

Year by year, as mining is being carried out at successively deeper levels, more and more problems arise. For instance, much more water seeps into the mine workings, and must be pumped out against higher heads. The ore must be hoisted a greater vertical distance, and therefore it takes more hours per day to hoist a given tonnage. Rock pressures become greater and the stability of the mine, now becoming more and more honeycombed with old workings, becomes a serious matter. Rock temperatures also increase with depth, and this problem, together with the increased airflow resistance offered by a deeper mine, calls for greater quantities of ventilating air to be circulated.

Some mines in the world, particularly those in South Africa, are very deep. In such mines, the ventilating air must be refrigerated (air-conditioned) so that comfortable conditions may be provided for the working miners.

2. Horizontal Bedded Deposits

By far the most important types of bedded deposits are coal seams.

Hypothetically, we are going to assume that our coal seam is six feet (2 m) thick, and extends in all directions horizontally to the boundaries of the mining property. The sedimentary rock above the seam is usually of shale or sandstone and is termed 'roof rock'. Similarly, the rock below the seam is called 'floor rock' (see Fig. 14).

If our seam occurred in mountainous country, then we could gain access to it by an adit, in which case it would be called a 'drift mine'.

But if, as we assume, the surface terrain is roughly horizontal, we will need to sink two shafts from the surface through the seam. One of these would be a main working shaft, and the other a ventilation shaft.

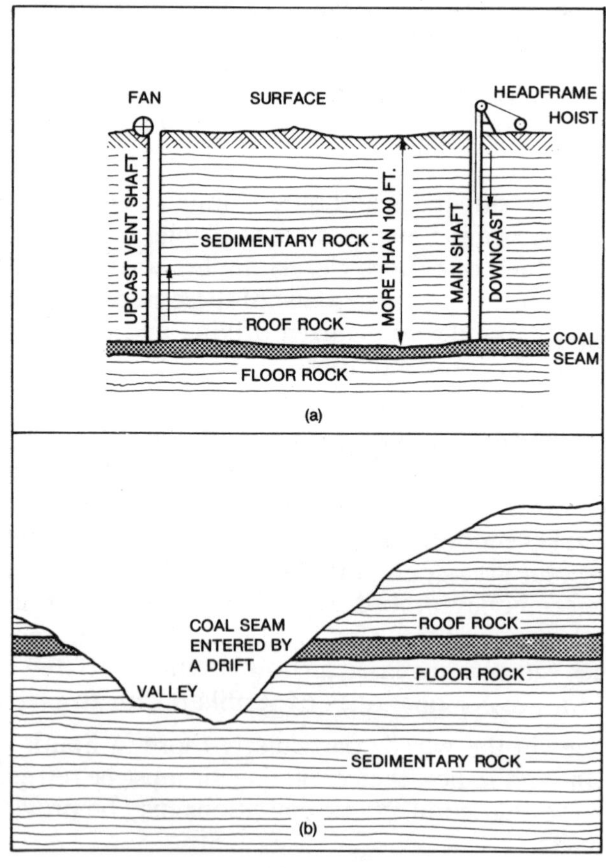

Fig. 14:
Hypothetical Underground Coal Mine Development
(a) Shaft Mine in Flat Country
(b) Drift Mine in Hilly Country

Where the shaft passes vertically through the coal seam, we leave a shaft pillar of unmined coal to protect the shaft and to maintain stability. For this reason, we sometimes sink both shafts in the same general area so as to use the same, though perhaps a somewhat larger, shaft pillar.

Apart from these two shafts which pass through the overlying sedimentary rock, all other workings will be within the seam of coal, so that all excavations made will incidentally produce coal.

In order to develop the mine, our huge slab of coal, in this case six feet (2 m) thick and extending horizontally to the property boundaries, will be dissected into sections or 'districts' by extending main entries from the shaft pillar to the boundaries in each direction. The mine is now ready for the systematic removal of coal, by one of three main methods.

(1) The Room-and-Pillar Method: Here, the coal is extracted by driving 'rooms' about 18 to 20 feet (6 m) wide on about 60 feet (20 m) centres. This means that a pillar of coal about 40 feet (14 m) wide is left intact between adjacent rooms. After proceeding in the rooms for about 60 or 80 feet (20 or 30 m), crosscuts are made at right angles to the general direction of mining. These crosscuts link up the rooms, and we get a rectangular pattern of unmined pillars about 40 by 60 feet (14 by 20 m) to support the roof rock (see Fig. 15). Meanwhile, all the work performed in driving rooms and crosscuts results in the production of coal.

These rooms and crosscuts are advanced by two alternative systems, (a) the **conventional mining system**, in which a slot called a 'kerf' is cut across the face of the coal at floor level; holes are drilled above this and charged with special 'permissible' explosives. When the holes are fired, the broken coal is loaded by a loading machine into transport vehicles to feed a belt conveyor in the main entry for haulage to the shaft; and (b) the **continuous mining system**, which involves the use of a highly integrated mobile machine called a 'continuous miner' (see Fig. 16). This machine mechanically rips coal out of the face as it keeps moving forward. The broken coal falls onto a steel apron that feeds a short in-built conveyor for transferring the coal to the rear of the machine. Here the broken coal is handled by one of a variety of methods to deliver it to the main conveyor system in the main entry. With either system, particular attention must be paid to the need to allay the dust, to ventilate the faces, to remove seepage

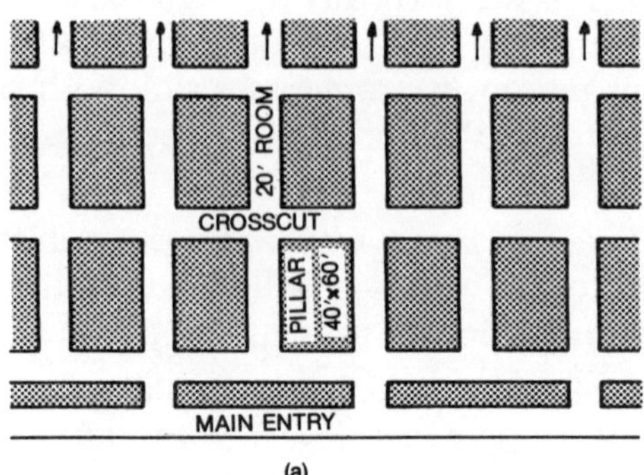

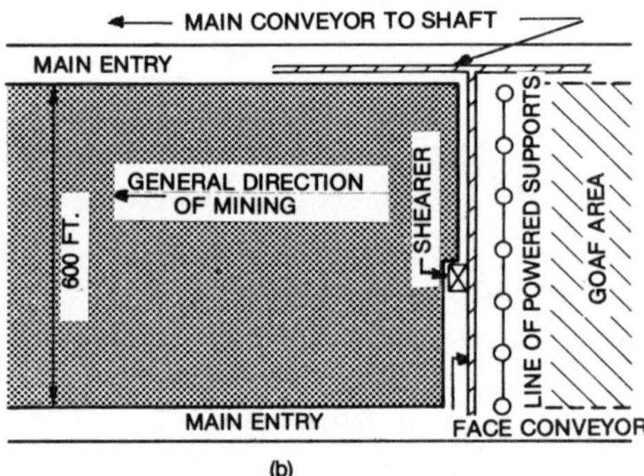

Fig. 15:
Two Methods of Mining Coal Underground
(a) Room-and-Pillar Method
(b) Longwall Method

water, and to give additional support to the roof, either by wooden props or with roofbolts.

One of the chief disadvantages of the room-and-pillar method is that most of the coal is left behind in the pillars. This would represent a national waste, but most companies attempt to recover the pillar coal by special methods at a later stage of mining. At least some of the pillar coal is recovered in this way.

(2) The Longwall Method: This has many advantages over the room-and-pillar method, chiefly because very little coal is left behind unmined in the pillars. Large panels of coal are initially developed,

Fig. 16:
A Continuous Miner

perhaps 600 to 1000 feet (200 to 300 metres) in width. In retreating from the boundary, the whole face is mined in slices, back towards the shaft. A shearer machine skims coal from the face and dumps it on a face conveyor belt which then delivers it to the permanent conveyor in the main entry. To support the roof near this long face, a line of specially designed cantilevered powered supports, like massive hydraulic jacks, is maintained in position. As the face advances, the face conveyor and the powered supports are moved forward in a regulated programme. Meanwhile, the roof rock is allowed to collapse behind the line of support in what is known as the 'gob' (or 'goaf') area.

(3) The Shortwall Method: This is a hybrid method developed in Australia. It involves the better features of the room-and-pillar and the longwall methods, in that a continuous miner operates under the protection of the type of cantilever roof support used in the longwall method.

The above descriptions of coal mine practice illustrate the main principles. Many complications can arise. For the sake of brevity and simplicity, these are purposely omitted.

When the softer types of industrial minerals, such as potash and trona, occur in bedded deposits, they are mined in the same general way as coal seams, using continuous miners in a room-and-pillar method, or otherwise, in the longwall method.

With hard industrial minerals and with metalliferous ores in bedded deposits, the face between the pillars needs to be drilled and blasted. In these cases, pillars are left in an irregular pattern, mainly only where (a) the roof rock is locally weak, or (b) the ore is of sub-marginal grade.

Surface Mining

Where a bedded type mineral deposit occurs at or near the surface, or where a wide lode exists with a large surface expression, it can usually be mined at a lower unit cost and with fewer problems by Surface Mining (see Fig. 4).

Coal seams lying within say 100 feet (30 m) of a reasonably level surface can be conveniently mined in this way. This is known as 'strip mining', because the superincumbent overburden is first stripped off and cast aside in order to expose the seam of coal in strips (see Fig. 17). After the coal is removed, the overburden is replaced from the adjacent new strip and the area is revegetated, using the topsoil previously skimmed off and stored separately.

A slightly different procedure is involved in mountainous country where the coal seam outcrops around a particular contour of a mountain. This is called 'contour strip mining'. As much of the overburden as possible is removed and stacked in a valley to expose coal until the resulting 'highwall' is likely to become dangerous. After mining the exposed coal around this contour, a further quantity of coal can be recovered from beneath the mountain by means of 'auger mining' (see later).

Strip mining is usually carried out on a large scale. For this purpose, gigantic machines such as walking dragline excavators, power shovels or bucket-wheel excavators are used to strip the overburden (see Figs. 18, 19, 20). The coal is then mined with smaller power shovels, loading into off-highway trucks. In most cases, a certain amount of drilling and blasting is required to loosen hard bands of overburden material, and therefore to facilitate the action of the excavators.

Bedded deposits of the softer industrial minerals, occurring near the surface, can be handled in the same general way.

The harder industrial minerals and the disseminated metalliferous or replacement deposits may also be recovered by surface mining, generally called 'open cut work'. These deposits are often no more than 40 feet (12 m) thick, under overburden of a maximum depth of say, 50 feet (15 m). In these cases we are dealing with hard rocky overburden and ore. Much more sophisticated methods must be used; these involve drilling and blasting, and a positive type of loading machine, such as a power shovel, is called for. Many of the porphyry copper mines in Arizona and New Mexico have adopted these procedures.

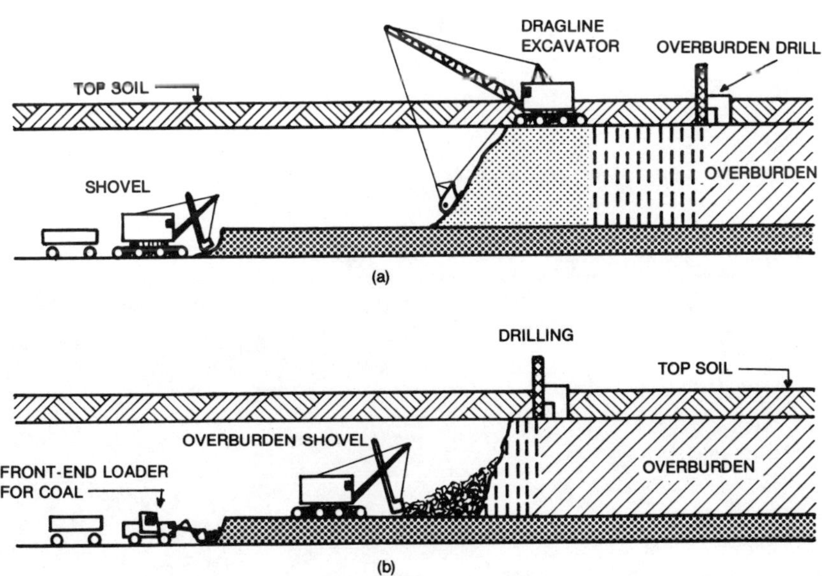

Fig. 17:
A Coal Strip Mine Operating
(a) by Dragline Excavator
(b) by Power Shovel

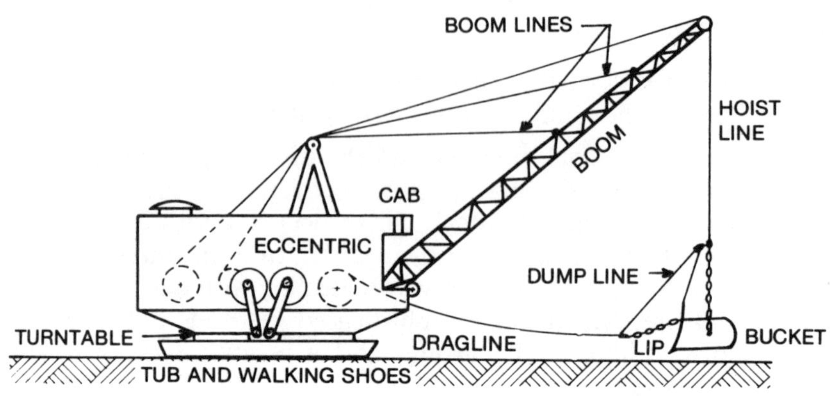

Fig. 18:
A Walking Dragline Excavator

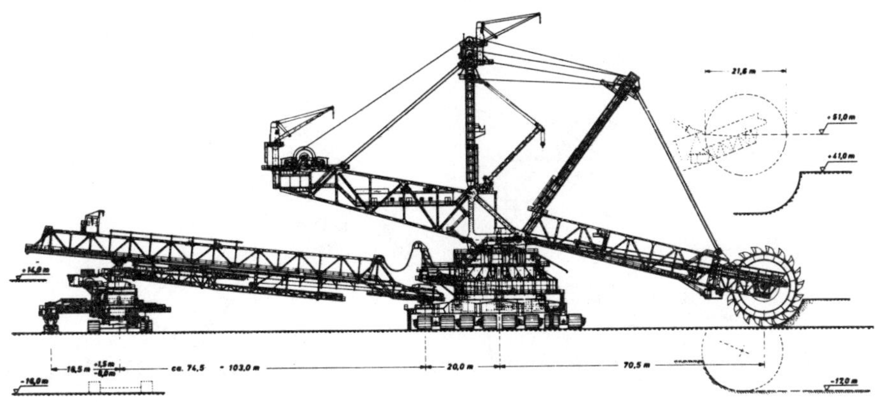

Fig. 19:
A Bucket Wheel Excavator

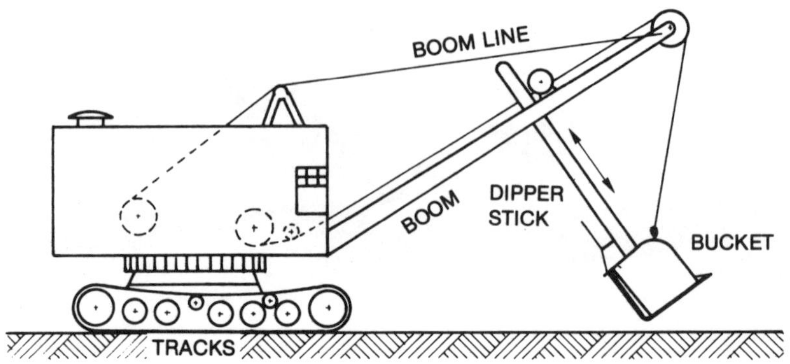

Fig. 20:
A Power Shovel

For steeply dipping wide lodes with a considerable vertical extent, additional considerations apply. After preliminary overburden removal to expose the whole of the orebody, a series of 'benches' is developed. These are stepped in heights limited to say 50 feet (15 m) for safety reasons. The horizontal part of the bench is designed to provide space to accommodate the broken ore from each blast, to deploy drilling machines to drill blastholes for blasting on the next lower bench, to allow space for a mobile power shovel to move in and load the ore into transport vehicles, and to provide a roadway for the latter. The general set-up is shown in Fig. 21. However, as open cut mining proceeds to greater depths, the exposed hanging wall and the footwall would become a source of hazard unless these walls were stripped back (by a similar benching procedure) to provide a safe general slope angle to the bottom of the open cut. The waste rock mined in the walls is transported to a dump area, whereas the ore is taken to the crushing plant and the mill.

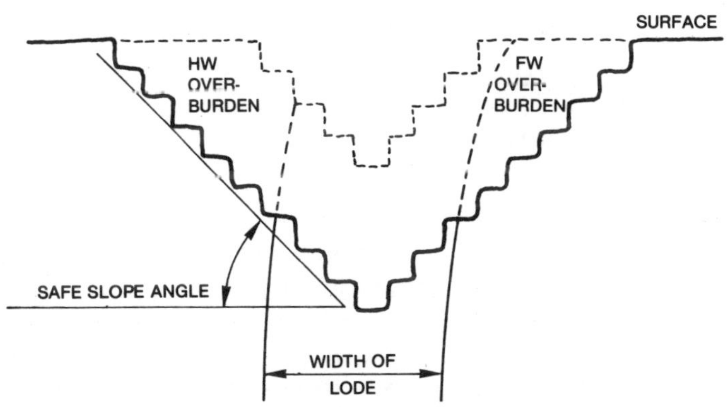

Fig. 21:
Open Cut Mining of a Wide Lode

As mining proceeds to deeper horizons, increasing tonnages of overburden are mined from the walls, adding considerably to working costs. A point is inevitably reached where it is better to cease open cut work and to continue to mine the deposit by converting it to an underground mining operation.

Quarrying is another type of surface mining operation, where rock is being mined (or 'quarried') for the production of 'construction stone', or building stone. Similar methods apply to those for open cut work in metalliferous deposits.

Another phase of surface mining relates to the harvesting of salt and other minerals from dry salt lakes or from lake brines as at Searle's Lake, California, and the Great Salt Lake in Utah. A number of minerals corresponding to sodium, potassium, and magnesium chlorides and sulphates (also lithium salts) are thereby produced by solar evaporation, fractional crystallization, and by mechanical scraping and loading of the respective materials.

Alluvial Mining

This is sometimes called 'placer mining'. It relates to the recovery of heavy (high specific gravity) minerals that have become concentrated in secondary deposits, following their separation by weathering agencies from primary veins higher up the watershed.

These dislodged mineral particles eventually become concentrated, typically in stream beds, below the gravels and immediately above the bedrock. Many collect in depressions in the top surface of the bedrock. Only the heaviest of minerals are concentrated in this way. Others are carried along the streams into lakes or to the oceans where they become concentrated in a different way by wave action, in beach sand or dune deposits.

The minerals more susceptible to concentration in alluvial deposits range from platinum (which when pure has a specific gravity of 21.4), and gold (19.3), through cassiterite (tin oxide), diamond, garnet, monazite, magnetite, zircon, to rutile (4.2). By contrast, the specific gravity of ordinary rock minerals, sand and gravel is 2.6. Some alluvial deposits are found in present stream beds. Others are found in terraces and in deep buried areas which were the beds of former streams. For this reason, although most placer mining activities are carried out as surface operations, others (such as 'deep leads') are mined by shallow underground methods.

Deposits described above are readily found and easily mined by relatively simple methods. Therefore most of the world's placer deposits (except those concealed by snow and muskeg in Siberia and a few other places) have already been found and worked out during the last 5,000 years or so.

The rich high-grade gold placers were mined simply by panning the gravels, or by shovelling the gravel into cradles or sluice boxes to shed the lighter material and to collect the gold behind riffles and similar traps. Mercury is sometimes used to collect fine particles of gold and silver by a process known as 'amalgamation'. The gold is later separated by retorting the amalgam, the mercury becoming vapourized and later saved by a condensing apparatus, and the resulting sponge gold is melted into bars.

Where gold-bearing gravels occur in the banks or terraces of streams, they can often be mined by sluicing away the bank material with an hydraulic monitor, using high pressure water. The gravels are then washed through large sluice boxes to collect the gold which settles behind the riffles. However, fine gold may travel along streams for many miles (km) before it reaches a wide river valley. Here the original stream loses much of its velocity and the fine gold is deposited in the slower moving stream in widespread gravel deposits. The gold content of these gravels is generally very low, but large scale dredging units are used to recover it at a profit.

Tin ore (cassiterite) is similarly recovered from river valleys by large integrated dredges in Malaysia, Thailand and Indonesia.

Heavy mineral sands found in beach strands, or in dunes above the beach, are also mined by hydraulic monitors or by dredging. Chief among these deposits are the beach sands of Australia from which rutile, ilmenite, zircon and monazite are recovered. These heavy beach sands are then treated in concentration plants to recover the minerals separately. The lighter clean silica sand is then returned to the beaches.

Ocean Mining relates to a new branch of the mining industry, recently under development to recover nodules of manganese dioxide from the sea floor, at depths ranging from 10,000 to 15,000 feet (3,000 to 5,000 m). The nodules are reported to contain metal values of manganese, copper, nickel, zinc, cobalt and molybdenum. Various methods such as dredging are currently being devised to recover these nodules and to deliver them to ocean-going barges for transfer to treatment plants on the shore. It is a new mining technique that is presently under development.

Another interesting aspect of ocean mining is presently under serious development as a new mining technology. This relates to the problem of recovering millions of tonnes of copper, zinc and silver from the hydrothermal mud deposits occurring at depths of 6,000 feet (2,000 metres) in the Red Sea. This problem is presently under investigation on behalf of the Saudi-Sudanese Red Sea Commission.

Non-Entry Mining

Under this heading, we have a number of widely differing approaches for recovering deep-seated minerals without mining them by conventional underground methods. This simply means that miners do not have to go underground to bring these minerals to the surface.

1. Oil wells

Crude petroleum and natural gas are brought to the surface by drilling oil 'wells' to depths ranging up to 20,000 feet (6,000 metres) to tap deep-seated oil deposits, occurring mainly in the interstices of relatively porous sandstone rocks. The oil is usually brought to the surface by pumping, aided by applied pressure. Oil well drilling rigs are highly specialized units, using tri-cone roller type non-coring bits in diameters up to 12 inches (30 cm).

2. Auger mining of coal

This procedure has already been explained. These large augers bore horizontally into coal seams exposed in the side of mountains, following the contour stripping operation. They are built in diameters up to six feet (2 m) and are capable of drilling and removing coal (like a giant carpenter's auger bit) from a depth of 200 feet (60 m) measured horizontally. The coal, as it emerges in the flights of the auger, is elevated into a bin for loading transport vehicles.

This is a remote procedure. No workmen need to enter the area from which the coal is extracted. It enables coal to be recovered under and behind the highwall without the need to strip overburden further.

3. Underground gasification of coal

Extensive trials have been made in Britain, the Soviet Union, the United States and other countries to gasify poor quality or thin coal seams *in situ* underground. It avoids the need for men to work underground, especially in these thin inaccessible seams. The principle is relatively simple but the applications in practice are difficult. The main components of the system are:

(a) an inlet shaft or borehole, to admit compressed air;
(b) a gasification area (combustion zone) where the seam is fractured by a jet of compressed air, and is ignited; and
(c) an outlet shaft or borehole to convey the product gases to the surface. Normally, a series of boreholes is needed to extend the combustion zone once the operation is commenced.

One of the main problems is that so far maximum efficiency is not obtained because of lack of control in the combustion zone. Nevertheless, important advantages are envisaged by the use of this method.

4. The Frasch Process for sulphur recovery

Sulphur-bearing strata are associated with buried salt domes in the Texas Gulf area. To avoid mining these deposits by conventional underground methods, Dr. Frasch invented a non-entry method in 1894.

Boreholes are drilled from the surface into the sulphur-bearing strata. Superheated water or steam is pumped down into these strata to melt the sulphur. The molten sulphur is forced to the surface in a central pipe by hot compressed air. In practice, many control problems can arise, but the method produces relatively pure sulphur.

5. Solution mining

In the old salt mines of Austria, water flows into the rock salt deposits, dissolves the salt and is pumped back as a brine solution to the surface for evaporation of the water and precipitation of the salt.

Underground deposits of copper or uranium ore may be leached *in situ* by percolating leach solutions through them, collecting the pregnant solution, and pumping it to the surface for precipitation of the copper or uranium salt. The successful operation of this method depends upon the permeability of the orebody. Various methods of pre-fracturing the orebody are under investigation.

Recommended reading:

Cassidy, S.M. ed. *Elements of Practical Coal Mining*. (New York: AIME) 1973.

Crawford, J.T. and Hustrulid, W.A. eds. *Open Pit Mine Planning and Design*. (New York: AIME) 1979.

Cummins, A.B. and Given, I.A. eds. *SME Mining Engineering Handbook*. 2 vols. (New York: AIME). Ch. 10, 12, 17, 21.

Peters, W.C. *Exploration and Mining Geology*. (New York: Wiley) 1978. pp. 175—208.

Pfleider, E.P. ed. *Surface Mining*. (New York: AIME) 1968.

Thomas, L.J. *An Introduction to Mining*. (Sydney: Methuen) 1978.

7. The Scope of Mining Activity

It is important to understand the great scope and complexity involved in the various stages of establishing and operating a mine from the grass roots until the ore deposit is completely mined out.

Naturally, these are much more complex in the case of the near-vertical type vein or lode deposits (which may call for the extension of the mine workings to great depths in extremely hard rock) than they are for the horizontal bedded type of deposit and for any type of surface mining. To cover the whole range of these activities, it is therefore considered necessary to describe the stages involved in the exploitation of such a metalliferous vein or lode type of deposit.

In the **Preliminary Stage**, by various techniques of prospecting and exploration, we locate a mineral deposit and stake the ground; or acquire the deposit under an option agreement or by direct purchase. Then we need to employ costly and sophisticated methods of diamond drilling and further underground exploration work to evaluate the form, shape, grade, and tonnage available in the deposit.

The next step is to prepare a feasibility report to provide basic evidence of the probability of mining the orebody over a period of years that will yield a profit commensurate with the risks involved. Such a feasibility report needs to be extremely comprehensive. It should incorporate the study of alternative mining methods with the associated degree of development planning, the projected working cost estimates of current development work; of stoping the ore; of crushing, grinding and concentrating the ore; of transport of concentrate to a smelter and the smelting costs; and of transport of basic metal products to a refinery or otherwise to market. All of these estimates need to be prepared in terms of detailed labour needs and costs, work efficiencies, costs of supplies, maintenance, power, water and overheads.

The overall working costs must then be related to the projected overall earnings year by year from sale of metal products based upon future metal prices (which are customarily quite variable), or upon long-term sales contracts that may be anticipated. In this way, a cash-flow table and an overall profit estimate can be determined.

Armed with this feasibility report, prepared and checked by competent mining engineers, the board of directors can then make a decision whether or not to form a company and proceed with the opening and operation of a mine.

If they make an affirmative decision, the **Development Stage** comes next. This is necessary to prepare the deposit for systematic mining by a well-planned framework of mine development, including the sinking and equipping of a shaft. Similarly, the metallurgical testing of a representative sample of the ore is carried out. Based upon the results secured from these tests, the design of a concentrating mill is prepared, materials and equipment are ordered, and erection of the concentrator proceeds. Similar procedures would be used if the erection of a smelting plant is also involved.

During the same period, and especially where there are no nearby towns or facilities adjacent to the mine site, an ample 'infrastructure' programme would need to be provided. This could include the installation of employee housing, workshops, powerhouse, water supply, amenities, health services, schools, roads, railway, airport and perhaps even a seaport.

Depending upon the scale of operations proposed, the above two stages of establishment could take up to five years of lead time and call for the capital expenditure of hundreds of millions of dollars, before a single ton (tonne) of metal can be produced or sold, or a single dollar earned.

The third stage is the **Production Stage**, during which actual mining of the deposit begins by using a particular method of stoping, mainly employing drilling, blasting and loading techniques in each stope. Associated activities include arrangements for sampling and surveying, pumping the seepage water out of the mine, providing adequate ventilation and lighting and other facilities, supporting the various excavations to prevent collapse, hauling the broken ore in ore trains along the various levels from the stopes to the shaft, hoisting the broken ore up the shaft to the surface and dumping it into ore bins, as well as a wide range of detailed minor activities.

From the surface orebins, the ore is drawn out and fed successively to crushers and grinding mills. The ore must be ground into extremely fine particles to separate the valuable minerals from the worthless rock materials (gangue etc.) by some method of concentration. The worthless gangue materials are discarded as tailings, most of which may be sent back underground to fill the empty stopes for support of the walls.

The mineral concentrate is then filtered and sent to the smelting plant where fluxing materials are added and the whole charge is roasted and smelted to produce (a) a worthless slag, and (b) ingots of the metal which can then be sold. Or alternatively, these ingots can be sent to a 'refinery' where the refined metal and perhaps several co-product metals are produced for the market.

The time involved from the breaking of the equivalent of a ton (tonne) of metal as ore in the stope to the receipt of the metal ingot by the buyer as a basis of sale may be of the order of six months. Consequently, a considerable amount of working capital is required.

Production then keeps going at a steady designed rate. Mine development work must keep ahead (perhaps two years ahead) of ore production activities. As the orebody becomes exhausted, more dollars are called for to explore for possible extensions of the orebody to provide more tonnage; or alternatively, to locate a new orebody elsewhere. The period during which an orebody becomes exhausted is equivalent to a planned extraction rate that yields a projected life of 20 to 25 years for an individual mine.

In the **Final Stage**, when the orebody becomes exhausted, mining and concentrating activities cease, the mine fills with water and is abandoned. Meanwhile, all equipment installed underground is usually left and written off (because the cost of recovery and reconditioning is too great), and very little of the surface plant is salable above scrap value, especially if the mine is in a remote locality.

The restoration and landscaping of the surface dumps, which had been proceeding actively during the mining operations, must now be completed and the various mine openings made safe. The company is liquidated unless it can acquire another orebody elsewhere.

The above simplified analysis will serve to demonstrate the great complexity of operations involved in the establishment and working of an underground multi-level metalliferous mine. As the mine workings become deeper with the process of time, the problems involved become significantly greater. There is probably no more complex operation in the whole gamut of industrial activity than that of a multi-level underground metal mine.

If the mine happens to be near a townsite with existing housing, power, water supply, communications and other facilities, then the infrastructure needs are greatly reduced.

Coal mines and others based upon bedded deposits generally involve fewer and simpler procedures.

Recommended reading:

PETERS, W.C. *Exploration and Mining Geology*. (New York: Wiley) 1978. Ch. 6, 7, 8.

8. Surveying and Mapping

Surveying is an important facet of mining operations, particularly where extensive underground workings are involved.

The geographical location of all underground openings can be determined only by surveying traverses plotted on a system of mine co-ordinates, because visual sightings of fixed monumental points are impossible as on the surface. The positions of all mine workings with respect to each other are of paramount importance.

The mine co-ordinate system is preferably based on the northeast quadrant so as to avoid negative readings. This means that the origin should be established well to the southwest 'corner' of the mining property, and the axes established due north and east (or at least, parallel and normal to the general strike of the orebody). Similarly, third dimensional readings should be established with the datum at the highest point of the mine property; or at sea level; or an even number of hundreds of feet (metres) above sea level.

A permanent base line with respect to the meridian is always established on the surface. This can be transferred to each level without much difficulty by using inclined sights wherever an inclined shaft is available. But for vertical shafts, the transfer of the meridian to each level underground is an operation where considerable skill and expertise is required.

The time-honoured method was to suspend steel piano wires down the shaft with heavy plumb bobs attached. Where two such shafts give access to the levels, one wire can be suspended down each shaft. The co-ordinates of each wire can be established at the surface, and therefore the distance between and the bearing of the line between them determined. By sighting back to these wires, a traverse can be run and the meridian established on each level. However, when only one vertical shaft is available, both wires must be suspended through this one shaft, and the much shorter base line creates greater difficulties and a general loss of precision.

Special methods have been developed to enhance the degree of precision. Instead of piano wires, laser beams have recently been introduced. Again, the gyro-theodolite has been developed to transfer the meridian underground.

Underground traversing is performed on each level by using a transit type theodolite set up at a station, to measure the horizontal angle between the stations on either side. The cross hairs of the instrument need to be illuminated, as do the horizontal and vertical circle scales.

Because of dubious floor conditions, it is customary to establish stations in the solid back (roof rock) of a drift or gallery. A wooden plug is driven into a short hole drilled into the rock and a brass spad is driven into the under side of the plug. A plumb bob is suspended from this spad. The plumb bob cord is illuminated by the chainman with his cap lamp at each sighting.

With each traverse line, offsets to the drift walls are measured at intervals with a tape. After the traverse is calculated (in co-ordinates) and plotted on the mine maps, the traverse lines are pencilled in and the offsets are used to plot the walls of the drift.

It is usual to use a permanent working plan (map) for each level, unless the orebody has a low angle of dip, in which case composite level plans may be maintained. At the end of each month, extensions to the working faces involve the establishment of new stations, and the new work is plotted on the plans. Cross-sectional drawings of the mine workings are also maintained at important intervals along the strike of the orebody; and also a longitudinal projection parallel to the strike.

Within the above framework of level traverses, small lateral extensions and stope work between levels can be plotted readily by the use of compass and tape.

Traverses, formerly laboriously calculated by latitude and departure methods to determine precise co-ordinates, are now more readily determined by a computer programme.

Mine surveying is one of the most complex applications of plane surveying, calling for a high degree of accuracy, often under difficult situations in confined spaces.

9. Drilling and Blasting

Soft materials, evaporites and some other industrial minerals, can be readily extracted from the deposit by specially designed mechanical excavating plant. But in some cases, where these are mined on the surface, the overburden needs to be loosened by blasting.

On the other hand, metalliferous ores and the harder industrial minerals and construction stone are much too tough, and can be broken into small pieces economically only by the use of explosives. This process of fracturing rock (breaking ground) involves the drilling of holes to accommodate the explosives which must be located within the rock mass.

Drilling

In general, these holes are drilled about 1 3/8 inches (3.5 cm) in diameter up to 10 feet (3 m) deep. Such hard rock is drilled by percussive methods, where the drill steel, with a tungsten carbide tipped chisel bit, is struck on the rear end with a succession of blows by a piston operating within the cylinder of a hammer drill under the action of compressed air at about 100 pounds per square inch (700 kPa) pressure. Such drilling machines are either hand-held (jackhammers) for vertical 'down' holes, or have special mountings for horizontal, or vertical 'up' holes. In these cases, they are fitted with an air feed leg (see Fig. 22).

These drilling machines, though portable, are connected to a compressed air pipe line through flexible hoses. To allay the dust caused by drilling the rock, a jet of water is passed through the machine and the hollow drill steel. Otherwise, the dust would cause a serious health problem, adversely affecting the lungs of the operator. Therefore,

Fig. 22 a:
Several Types of Pneumatic Rockdrills

Fig. 22 b, c:
Several Types of Pneumatic Rockdrills

these machines must also be connected with a water hose. Much larger trailer-mounted or self-propelled drills are used for surface mining. Depending upon the conditions at the ore or rock face, a particular pattern of holes must be drilled to make the best use of explosives.

Explosives

The use of explosives has contributed largely to our present high standard of living by minimizing the arduous nature of rock excavation work and by increasing the productivity of such operations.

Briefly, an explosive is a mixture of substances which, when initiated, produce gases which exert high pressure and thereby rupture the surrounding rock.

There are three main types of explosive used for rock breaking operations:

(1) **Low explosives,** such as Black Powder, consisting of sodium nitrate, powdered charcoal, and sulphur. This mixture, when ignited with a safety fuse, burns rapidly (deflagrates) and develops pressure from the gases formed in the reaction. It is now used only where a relatively low pressure is desired, such as in blasting building stone or monumental stone.

(2) **High explosives,** typically dynamites, made from nitroglycerine and ammonium nitrate. When initiated by a powerful shock from a 'detonator', such a charge detonates at a high reaction velocity and develops considerable heat and pressure.

(3) **Blasting agents,** made from ammonium nitrate and a reducing agent, are much safer to handle, are cheaper than high explosives, and in many cases, just as effective. When initiated by a shock from a detonator, they explode at a high velocity of detonation. There is a dry type (a mixture of ammonium nitrate and fuel oil, 94:6 by weight), but with minimal water resistance; and a slurry type, with high water resistance. These blasting agents are now used in most rock fracturing operations.

Safety fuse is a long flexible cord with a core of black powder. When set alight at one end it burns at a regulated rate so that the miner has time to retreat to a place of safety before the charge goes off.

A **detonator** is a small hollow aluminium shell, about 1/4 inch (6 mm) in diameter, closed at one end and containing a small charge of special initiating explosive pressed into the closed end of the shell. The rest of the shell will accommodate the end of a length of safety fuse which is crimped to the shell. In use, the detonator is inserted into a cartridge of high explosive or into a charge of blasting agent. As the safety fuse burns down, it ignites the detonator which then initiates the main explosive charge with a powerful shock.

Alternatively, detonators may be provided with an integral electric firing mechanism instead of safety fuse. When ready to fire, an electric current is passed (from a safe place) through connecting wires, and the detonator sets off the main charge. These are termed 'electric detonators'.

There are many other accessories available for initiating an explosive charge.

To charge a blast hole, high explosives in cartridged form are inserted with a wooden tamping rod. One of these cartridges carries the detonator. Alternatively, a blasting agent may be introduced into the hole in bulk by a pneumatic charging machine. A detonator unit is also added at the appropriate position. The charge is then sealed with stemming material (to prevent the escape of the gases formed in the explosive reaction) and is then ready to fire. A particular blast may embrace many such charged holes.

Open cut work and quarrying call for similar procedures in large scale applications of drilling and blasting.

Recommended reading:

Gregory, C. E. *Explosives for North American Engineers* 2nd edition. (Clausthal-Zellerfeld: Trans Tech Publications). 1979.

10. Mine Support

One of the more obvious aspects of mining technology is the need to ensure that the underground workings do not collapse. If this were allowed to happen, there could be great losses of human life and much suffering, considerable damage to equipment, and significant dislocation of the whole mining operation. It is therefore of paramount importance to maintain the stability of the rocks enclosing the mine workings.

The stresses in virgin rock are due mainly to pressure from the weight of the superincumbent strata. As soon as development openings are made in the rock, the stress pattern alters. This depends, with elastic rocks, on the cross-sectional shape of the opening. Because elastic rocks are brittle, they tend to fracture under increased stress. It has been observed that around any opening there is a fractured zone of relatively loose rock, behaving inelastically; and some distance beyond this zone there is a zone of high stress.

The rocks from this fracture zone (or clastic zone) tend to fall or move into the opening, but they are restrained to some extent by compressive and frictional forces. Movement inwards can be further restrained by spraying with a mortar such as 'gunite' or 'shotcrete'; otherwise some other medium of support must be used.

Gunite is a mortar of cement and fine sand (1:3 to 1:5), dry-mixed and loaded into a cement gun and blown through a nozzle by compressed air mixed with water. Shotcrete contains coarse aggregate ($^1/_4$ to $1\,^1/_4$ inches, 6 to 30 mm), sand, and cement, with chemical accelerators. Shotcrete can be placed dry or by a wet process, using a compressed air nozzle. The effect of these mortars is to bind the rock and to distribute the stresses across any discontinuity in this clastic zone.

For narrow openings, the effect of rock falls from the back can be minimized by arching. In this case, the main superincumbent load is re-distributed to the side walls.

Otherwise, the flow of loose rock into an opening can be inhibited by timbering, steelwork, concrete lining, or rockbolts. In stopes, the walls may be supported by pillars, by timbering, rockbolting, or filling with waste rock or mill tailing.

Pillars

Pillars (generally of low grade ore) are left unmined at intervals to support the walls in some methods of stoping. Otherwise, this is a regular procedure in room-and-pillar mining (of coal and some industrial minerals) in bedded formations. See Figure 15a.

Timbering

The use of timber has been traditionally used for thousands of years to support mine openings. The timber must be wedged firmly to the rock, perpendicular to the grain. It is used in the following ways:

(1) As stulls, between the walls of stopes (Fig. 11).
(2) As pit props, in bedded deposits.
(3) As cribs, to support a weak portion of a stope back (Fig. 23).
(4) As chute cribbing, in cut-and-fill stopes (Fig. 13).
(5) As pre-cut modular units, in square set stopes (Figs. 24, 25).
(6) As pre-cut shaft linings (Fig. 5).
(7) In miscellaneous ways.

Steelwork

Steel sets are used in more permanent locations, such as at shaft stations. However, it is necessary to block the contact points (between steel and rock) with timber pads, to cushion the initial rock movement.

Where heavy ground is encountered in drifts, yieldable (non-rigid) arch steel sets are used. These allow the wall rock to expand slightly until it stabilizes, without distorting the set. The sets are made of special channel sections, held by bolted brackets, which nevertheless permit sliding of one section over another as the load builds up (Figs. 26, 27).

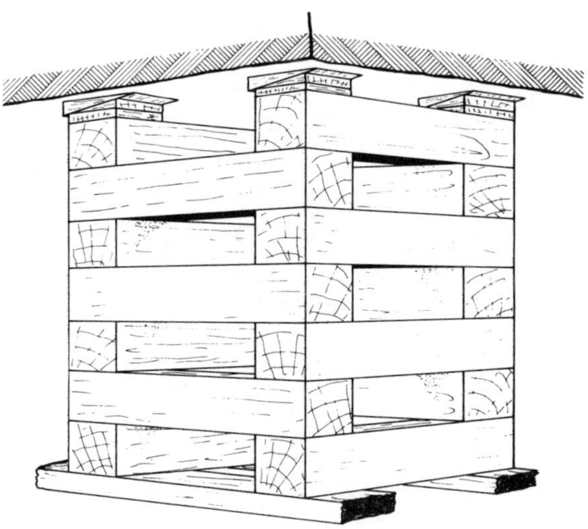

Fig. 23:
Timbered Crib for Ground Support

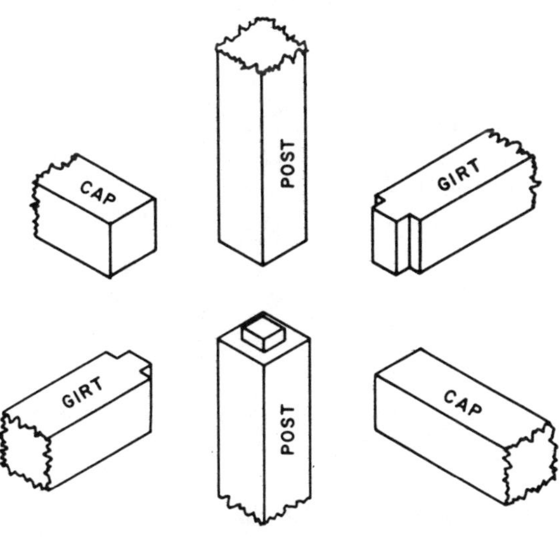

Fig. 24:
Details of Square Set Timbers

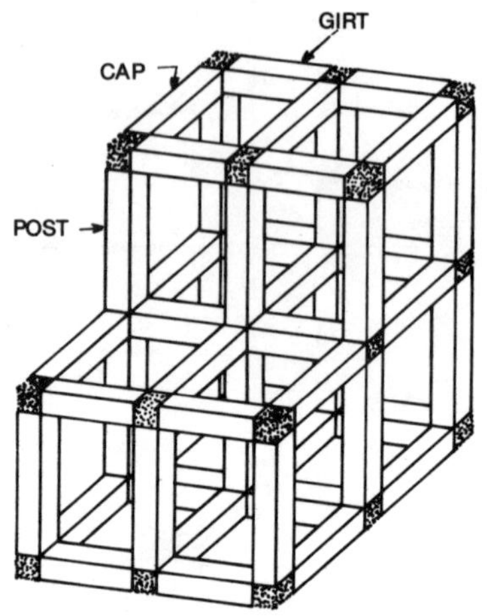

ASSEMBLY OF SQUARE SET MODULES

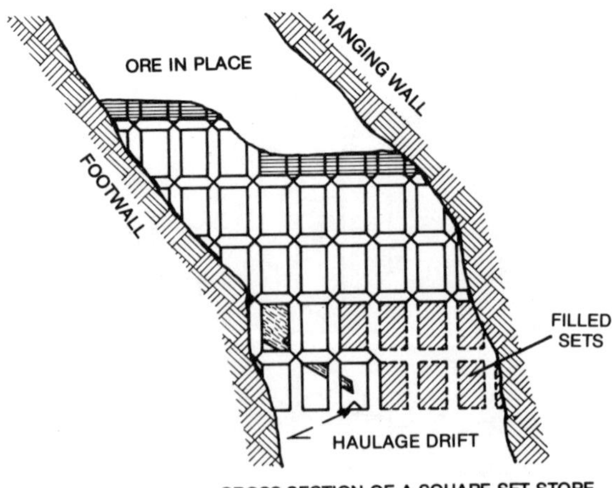

CROSS-SECTION OF A SQUARE SET STOPE

Fig. 25:
Square Set Timbering Arrangements

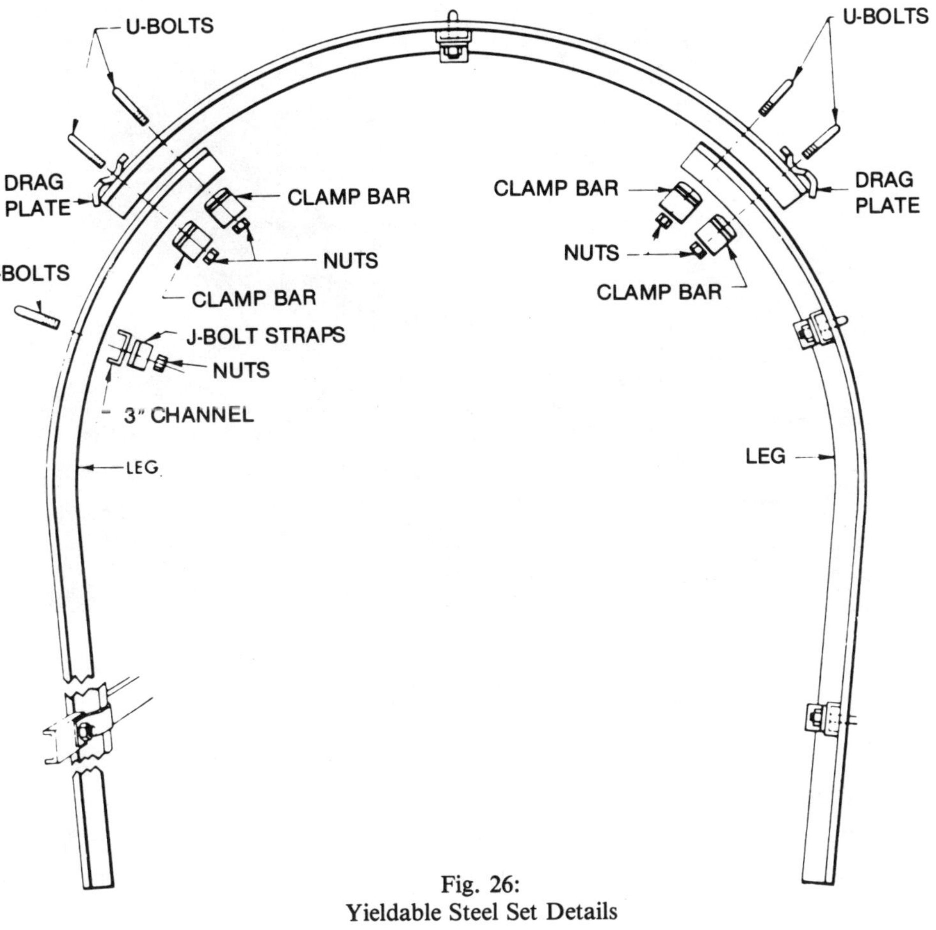

Fig. 26:
Yieldable Steel Set Details

Fig. 27:
Yieldable Steel Sets in a Drift

Concrete

Concrete lining is used mainly for tunnels and in shafts, usually under conditions set by contract specifications. Otherwise, concrete mortars are used as surface sprays for rock surfaces, such as gunite and shotcrete.

Rockbolts

These were designed as a sort of wall rock reinforcement to make rocks self-supporting. See Figure 28. Holes are drilled through the clastic zone into the solid rock behind. The bolts are inserted in a way that anchors them to the bottom of the hole. The nut is then tightened (with a torque wrench), thereby placing the bolt in tension and the rock in compression. Various types of bolt anchorage have been devised. A pattern of bolts is placed in any particular area, connected by wire mesh sheets, if rock fragments cause problems.

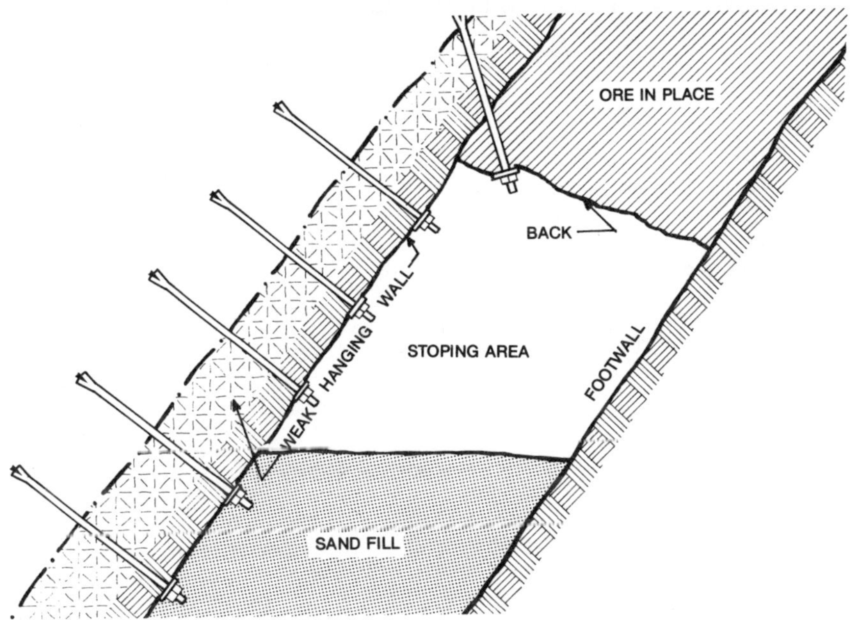

Fig. 28:
Rockbolts

Stope Filling

Where stoped out areas need support, the open cavity is customarily filled with waste rock or surface sand or mill tailing, placed manually or by pneumatic jets or hydraulically. Hydraulic filling of mill tailings is now a common practice although there is the problem of removing the excess water from the mine. In such cut-and-fill stopes, the surface of the wet fill can be consolidated more quickly by an addition of cement to the final run of filling at each cycle. This facilitates the passage of mechanized equipment over the new stope floor without undue delay.

Rock Mechanics Principles

In recent years, the science of rock mechanics has been developed by rigorous analytical methods. It is therefore an important aspect of mining engineering design. However, great care is necessary in applying laboratory and theoretical concepts to meet rock problems in the field because rock is seldom homogeneous or isotropic.

Recommended reading:

LAMA, R.D. and VUTUKURI, V.S. *Handbook on Mechanical Properties of Rocks, Vol. I—IV*. (Clausthal-Zellerfeld: Trans Tech Publications). 1974/1978.

ROBERTS, A. *Geotechnology*. (New York: Pergamon). 1977.

11. Haulage

The term *haulage* is applied to the horizontal transportation of ore and waste rock from stopes and development headings, generally along levels to the shaft, where it is *hoisted* to the surface. It therefore represents the horizontal component of the materials handling aspect of transferring broken ore from the vein underground to the surface.

Actually, three phases of haulage are involved: loading, transportation, and dumping.

Loading refers to the loading of broken ore at stope faces by LHD (load/haul/dump) mobile units (Fig. 29); in loading ore cars from chutes and draw points on the level (Figs. 9, 30); in loading ore or waste rock from development headings into cars; in loading coal from room-and-pillar faces into shuttle cars (Figs. 15a, 31); and the like.

Transportation is of two types:

(a) Short-haul transportation involves moving of broken ore from stope faces into chutes or nearby ore passes by LHD type units or scrapers (Figs. 29, 32); scraping ore from draw points in scram drifts to ore passes (Fig. 33); moving coal from the face to the main line conveyor (by face conveyors in the longwall operation, or by shuttle cars in the room-and-pillar method, as shown in Figs. 15, 31); 'gathering' of individual ore cars in a productive area by battery loco to marshal a full train rake; and such like.

(b) Main line transportation involves the moving of ore for long distances, such as by the use of main line conveyor belts which receive coal at various points from face conveyors and shuttle cars, conveying it to the main shaft, or up an incline (less than about 20 degrees) to the surface; or electric (or diesel) locomotive transport of ore trains from marshalling areas to the shaft.

Fig. 29:
A Wagner Scooptram

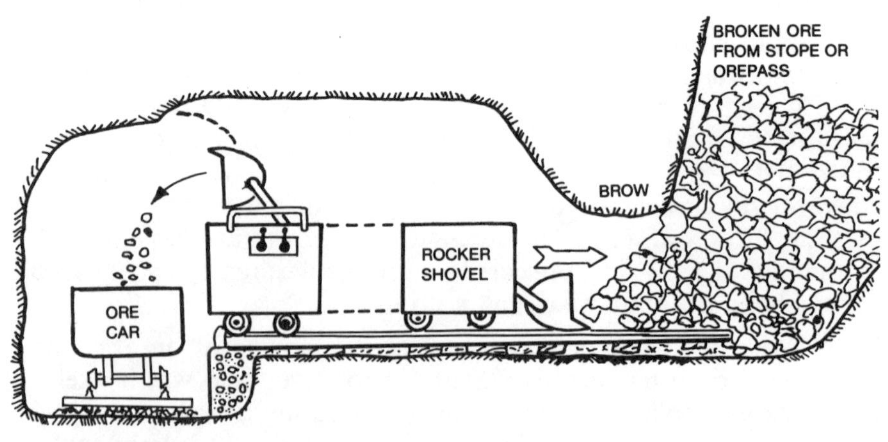

Fig. 30:
Loading Ore Cars from Draw Points

Fig. 31:
A Shuttle Car

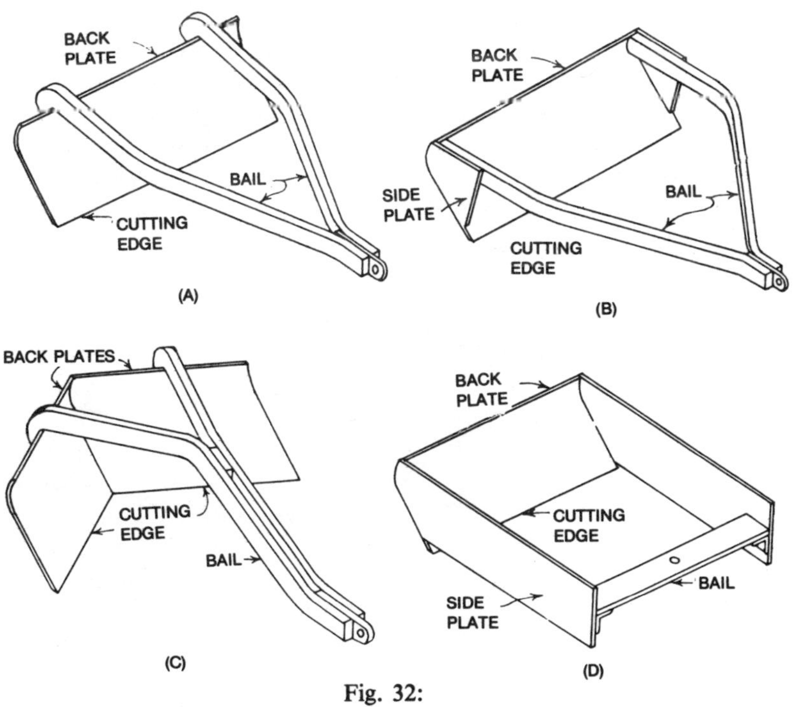

Fig. 32:
Scraper Units

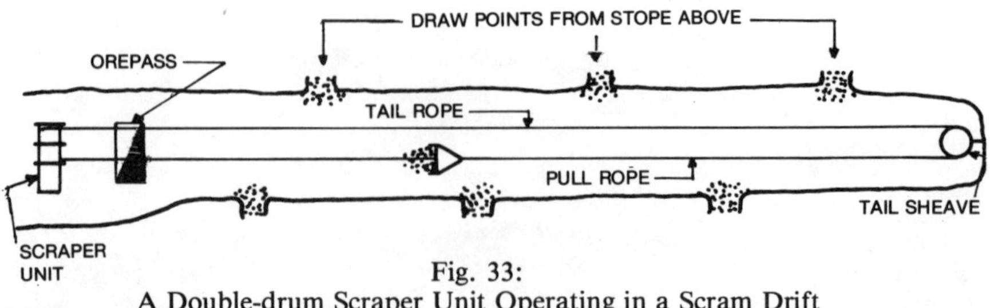

Fig. 33:
A Double-drum Scraper Unit Operating in a Scram Drift

For small mines, manual tramming of 1-ton (1-tonne) box type end-dump ore cars (Fig. 34) involves loading of individual cars at stope chutes (Fig. 9), pushing them along the level to the shaft into the cages (Fig. 38) for hoisting directly to the surface. In some older mines, animal transport is employed. A horse or mule pulls a rake of 2-ton (2-tonne) cars to the shaft, or to the surface orebins through adits where there is no shaft. In this latter case, underground stables are not required; nor is the daily hoisting and lowering of animals.

Dumping refers to the dumping of ore in stopes into ore chutes or ore-passes by LHD units or scrapers; and by scrapers in scram drifts into an ore-pass; of coal into a hopper over the main line conveyor belt by shuttle cars; of ore or waste rock from a development heading into cars by small rocker shovels or LHD units; of ore at the discharge point near the shaft (on particular levels) through a grizzly into an orepocket, which stores ore for loading, with or without primary crushing, into skips to be hoisted to the surface.

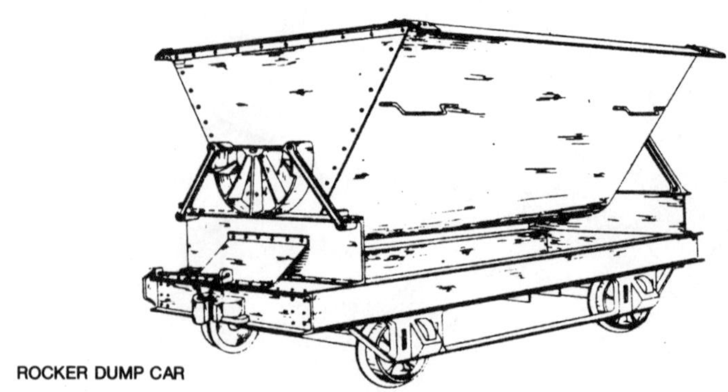

Fig. 34:
Various Types of Ore Cars

GRANBY TYPE OF CAR

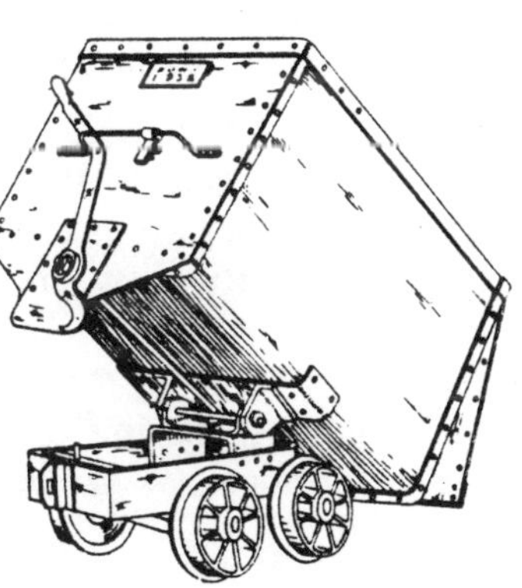

END-DUMP CAR

GABLE-BOTTOM CAR

Chutes and Drawpoints

The flow of broken ore under gravitational conditions from stopes is confined to chutes, orepasses and draw-points, from which it is generally loaded into ore cars.

Within cut-and-fill stopes, the chute is progressively built up through the filling by chute timbering (Fig. 13) or steel plate sections. At the bottom of each chute, where it meets the level beneath, the chute is fitted with a gate. When the chute gate is opened, the broken ore flows into and is allowed to fill the car (Fig. 9).

In the case of the draw-point, the expense of constructing and maintaining the chute gate is avoided because the ore flows through to the floor of a small crosscut off the haulage drift. The spread of ore on the floor is limited by the brow of the ore pass and the natural angle of repose (see Fig. 30).

Load-Haul-Dump Machines

These are specially designed mobile tyre-mounted vehicles for all three operations (loading, short-hauling and dumping) in confined spaces such as in mechanized cut-and-fill stopes.

Typical of the LHD machine range is the Wagner Scoop-tram (see Fig. 29). Because of their size and the need for maximum mobility, LHD machines are used in large scale mass production underground mining operations where access from the surface and from level to level is specially provided by a spiral or zigzag rampway. They are powered mainly by diesel engines fitted with exhaust scrubbers, but additional ventilating air is also required. Haulage grades up to 1 in 9 are economical, as are haul distances up to 500 feet (150 metres).

Scrapers

Scraper units are used mainly in short-haul operations, in stopes or scram drifts. The scraper proper is caused to be filled with ore and moved to a chute or orepass, under the control of a steel wire rope driven by an electric winch, having one, two or three rope drums (see Fig. 32).

With a single drum unit, the loaded scraper is drawn forward by the winch, but it must be moved back manually.

With a two-drum unit, one rope draws the loaded scraper forward; and a tail rope, passing through an anchored tail sheave, returns it after dumping to the ore pile (see Fig. 33).

In wide stopes, a three-drum unit can be used to adjust the direction of pull.

Scraper units are built in sizes from 10 to 200 horse-power.

Shuttle Cars

With the room-and-pillar method of mining coal or soft bedded deposits, the **shuttle car** has become the popular transport medium for carrying coal from the continuous miner at the face to the main line conveyor belt.

The bed of the machine is fitted with a short conveyor onto one end of which the broken coal is loaded. This conveyor moves the coal forward until the unit is fully loaded. It is then driven out to the nearest conveyor loading hopper and dumped before returning for the next load. Meanwhile, a second shuttle car is being loaded at the face.

Most shuttle cars are powered with AC motors by means of trailing cables. They drive and steer on four rubber-tyred wheels (see Fig. 31).

Locomotive Haulage

For large scale metalliferous mines, ore is usually hauled along levels to the shaft by trains of ore cars drawn by a locomotive, trolley-electric or diesel powered.

For short haul purposes, such as in train marshalling from chutes (gathering duty), small battery-electric locomotives are used. In medium scale mines, battery locos also perform main line duty. Battery units are changed each shift for re-charging purposes.

Electric trolley systems need an overhead trolley wire and bonded rails. Locomotive grades should not exceed three per cent. It is usual to provide for one-half per cent grades downhill for loaded trains.

For smaller mines, the track gauge is normally 18 to 24 inches (45 to 60 cm); and for medium to large mines, 24 to 36 inches (60 to 90 cm). For main lines, tracks need to be well designed and well maintained, with heavy rails (40 to 60 lb/yd, or 20 to 30 kg/m), good quality ties (sleepers) about 6x8 inches (15x20 cm) at 18 inches (45 cm) spacing, and ample ballasting clear of water.

Short radius curves are permissible in small mines; but for main line work in large mines, the radius should not be less than 75 feet (25 m).

For safe operation on main lines, an effective signalling and intercommunication system is necessary.

In all rail systems, the tractive effort of the locomotive exerted at the rim of the driving wheels is derived from the adhesion of the locomotive wheel tyres to the rails, and depends upon the weight of the loco and the condition of the tyres and track.

The Coefficient of Level Track Adhesion A is usually taken as 25 per cent for steel-tyred wheels on clean dry well-ballasted track. This means that the *Tractive Effort* developed (available) in lb is $T_a = 2000 \times AL$ (or 1000 AL kg), which is 0.25 times the weight of the locomotive (L tons or L tonnes) supported by its driving wheels.

The force exerted by a locomotive in drawing a train is known as the *Drawbar Pull D*. The drawbar pull available (D_a) for any loco is a function of the tractive effort developed minus the *Tractive Resistance* (R_l) of the loco itself. Tractive resistance refers to the rolling friction of a loco, or train, on level straight track. For roller bearings, the tractive resistance is usually calculated as 20 lb per ton (10 kg per tonne) weight; up to 30 or 35 lb per ton (15 or 18 kg per tonne) for plain bearings and poor track conditions. So we may take $R_l = 20 \times L$ lb $= 10 \times L$ kg.

This means that the drawbar pull available is

$$D_a = T_a - R_l = L\,(2000\text{ A} - 20)\text{ lb}$$
$$= L\,(1000\text{ A} - 10)\text{ kg.}$$

The *Grade Resistance* is usually taken as 20 lb/ton (10 kg/tonne) for each percentage of grade, plus or minus.

Curve Resistance and the force required to accelerate a train are customarily neglected in routine calculations.

The weight of locomotive required for a certain duty is derived from the tractive resistance of a train of known weight and the ruling grade.

The tractive effort required to move a loaded train (neglecting curve resistance and acceleration) is

$$T_r = L\,(R_l + 20\,G) + W\,(R_w + 20\,G)\text{ lb}$$
$$= L\,(R_l + 10\,G) + W\,(R_w + 10\,G)\text{ kg}$$

and the tractive effort available is $T_a = 2000 \times AL$ lb $= 1000 \times AL$ kg,

where L is the weight of the loco, tons (tonnes)
R_l is the tractive resistance of the loco, lb (kg)
W is the weight of the loaded cars, tons (tonnes)
and R_w is the tractive resistance of the cars, lb (kg).

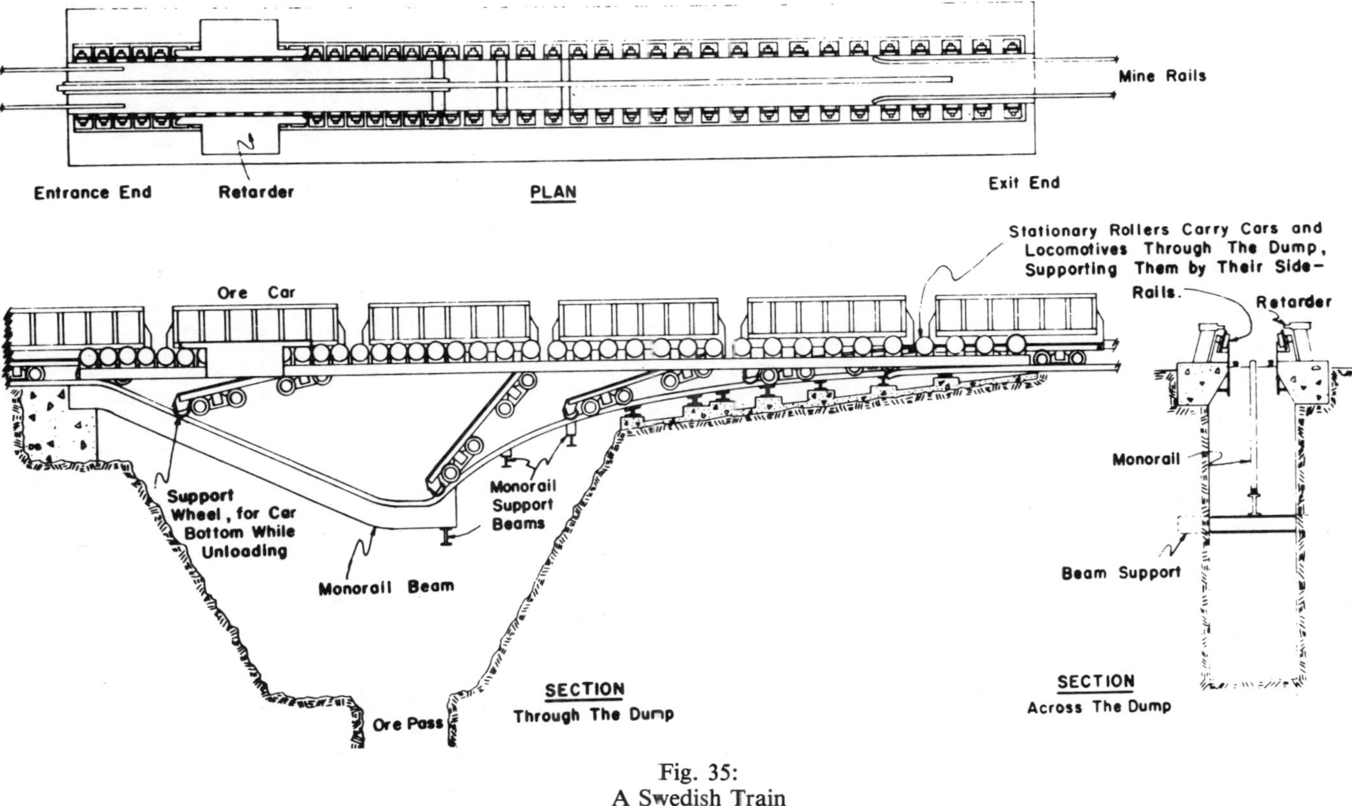

Fig. 35:
A Swedish Train

By equating these, the weight of the loco required is

$$L = \frac{W(R_w + 20G)}{2000\,A - (R_l + 20G)} \text{ short tons.}$$

$$= \frac{W(R_w + 10G)}{1000\,A - (R_l + 10G)} \text{ tonnes.}$$

For level track conditions,

$$L = WR_w/(2000\,A - R_l) \text{ short tons}$$
$$= WR_w/(1000\,A - R_l) \text{ tonnes.}$$

More sophisticated relationships can be derived from these basic equations.

Mine Ore Cars

These are of several different styles of construction (see Fig. 34). For manual propulsion, box type 1-ton (1-tonne) end-dump cars are designed to fit the hoisting cage in small mines.

When animal transport is used in adit haulage, side-tipping rocker-dump cars of around 2-tons (2-tonnes) capacity are customary.

For loco haulage, there is a choice between a gable-bottom discharge car, a granby side-dumper (using a jockey wheel on an inclined ramp), or a Swedish train (Fig. 35) where the complete car bottom, hinged at one end or side, falls away to ride on a monorail beam, supported by a guide roller.

Conveyor Belts

Conveyor belts are being employed on an increasing scale for underground transport, especially for coal. They are particularly adaptable for shallow bedded deposits, where the main belt can be extended directly to the surface, either through an adit or an inclined haulageway or slope.

Rubberized fabric troughed belts from 24 to 60 inches (60 to 150 cm) in width, running at 500 to 650 feet (150 to 200 m) per minute, can handle up to 4000 tons (tonnes) per hour, at slope angles up to 20 degrees.

12. Hoisting

The vertical component of the materials handling problem of moving ore to the surface is known as **hoisting**.

Early simple methods of hoisting, by man or animal power, included the windlass, the whip, or the whim (see Fig. 36).

Hoisting through vertical shafts is accomplished by the movement of an ore skip or cage 'conveyance', attached to a stranded steel wire rope which is caused to move by a motor-driven winch or hoist (winding engine). In order to avoid frictional forces between the conveyance (cage or skip) and shaft lining, the rope must be maintained in a dead central position in the compartment. This means that the rope direction must be changed above the surface to allow it to be driven by the winch, mounted on ground foundations. This change of direction is accomplished by means of a rope sheave mounted some distance above the shaft collar, and necessarily supported by a headframe, built of wood, steel or concrete (Fig. 7).

To drive the rope, the hoist is equipped with a drum to which the rope is anchored and on which the surplus rope is stored. Drums were formerly designed in various shapes, such as cylindrical, conical, cylindro-conical and bi-cylindro-conical, but now cylindrical drums only are used.

The simple hoisting system consists of a conveyance, a rope, a headframe sheave and a winch. But, except for small units, these are uneconomical in power consumption and lack adequate brake control. It is now general practice to balance the forces as much as possible, to avoid or reduce these two problems. This is achieved by a twin system, using two conveyances, two ropes (in two compartments), two headframe sheaves and a double-drum hoist. The two ropes are caused to move in opposite directions, so that a loaded skip

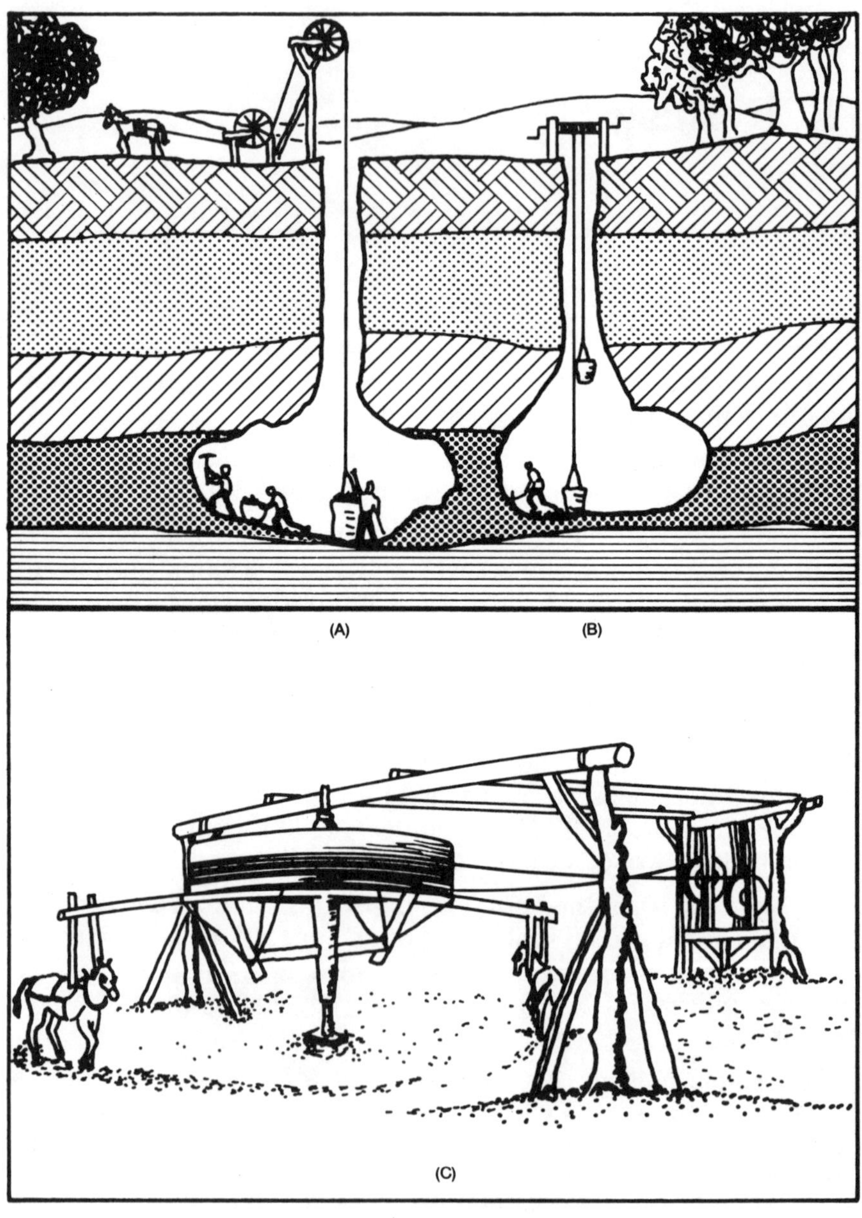

Fig. 36:
Simple Hoisting Mechanisms
(A) Whip; (B) Windlass; (C) Whim

being hoisted is partly balanced by an unloaded skip being lowered. This is the standard arrangement. Where cages are involved, one large capacity cage is operated with a counterweight in partial balance on the other rope. However, in small mines, men are lowered at the beginning of a shift in cages (about 9 men to a cage), after which the cages are replaced by skips for ore hoisting until near the end of the shift when cages are then used again to hoist the men. In some other mines, combination skip-cages are used throughout the shift, but this has several disadvantages.

The **hoisting cycle** involves four different periods:

(a) Acceleration period, from a skip at rest to full speed.
(b) Cruising period, in which the skip travels at constant speed.
(c) Deceleration and braking period, to bring the skip to rest.
(d) Rest period, during which one skip is being loaded (at one of the orepockets underground) while the other is being dumped automatically into the surface orebin.

Although some hoists run at speeds up to 3600 feet (1200 m) per minute, it is better to design for lower speeds (say, 2400 ft/min, 800 m/min) to reduce shaft maintenance costs.

When inclined shafts are used, the skips (although pulled by ropes) run on rails, and therefore lower hoisting speeds are necessary to avoid derailments and other mishaps.

Hoisting ropes are of stranded steel wire construction in various standardized forms and grades of steel wire.

In practice, a hoist rope is subjected to tension, compression, torsion, bending, impact, friction, abrasion and corrosion. This means that a mine hoist rope must be designed, manufactured and operated with due care to withstand these forces. To give adequate flexibility, a rope is made in strands each of a number of wires formed in a certain pattern.

The usual construction is of six strands formed around a central core of fibre. Each strand may have 7, 19 or 37 wires in increasing order of flexibility. In an ordinary lay, the wires are twisted in a direction opposite to that of the strands; with a Lang lay, in the same direction, thereby offering a greater continuous surface to abrasion, but a less stable rope (Fig. 37).

To reduce the effect of bending stress, the diameter of the rope, for any particular breaking strength, is important. From experience, the

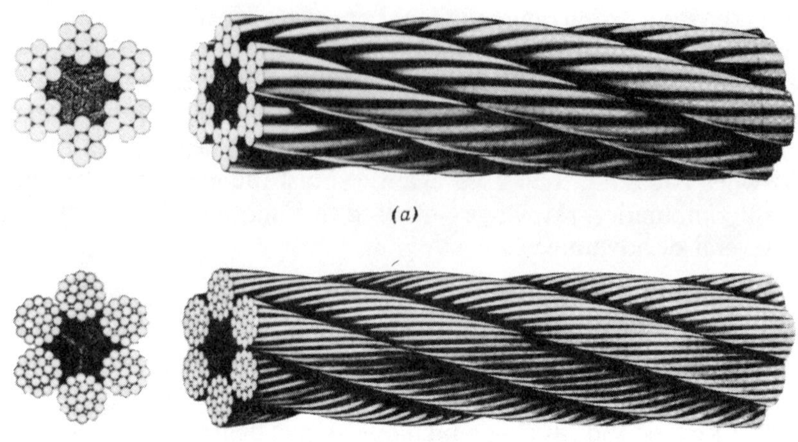

Fig. 37:
Examples of Ordinary Lay Hoist Ropes
(a) 6/7; (b) 6/19

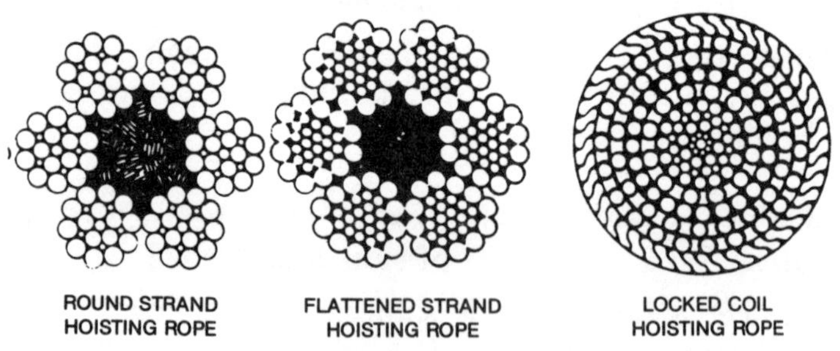

Fig. 38:
Various Steel Wire Rope Sections

rope diameter should be related to the diameter of sheave or drum around which it must be turned. For instance, a 6 x 7 rope of diameter *d* inches (cm) should not be used with a sheave or drum diameter smaller than 125*d*; for 6 x 19, 100*d*; and for 6 x 37, 80*d*.

In order to keep *d* to a minimum for any rope strength, the quantity of steel in a rope cross-section can be increased by using 'flattened strands', instead of the normal round strands; and still more, by using a 'locked coil' rope (see Fig. 38). These two latter developments give rise to hoist designs with smaller diameter drums and headframe sheaves, resulting in reduced inertia losses and lower operating costs.

This tendency to reduce inertia losses in the rotating mechanisms is still further enhanced by the development of the **friction hoist,** where a single rope is attached to the draw-gear of two skips, and passes over two headframe sheaves and around a drive sheave (instead of two drums) on the hoist (see Fig. 39). The rope is driven by the friction developed as it passes around an arc of the drive sheave (Fig. 40). The

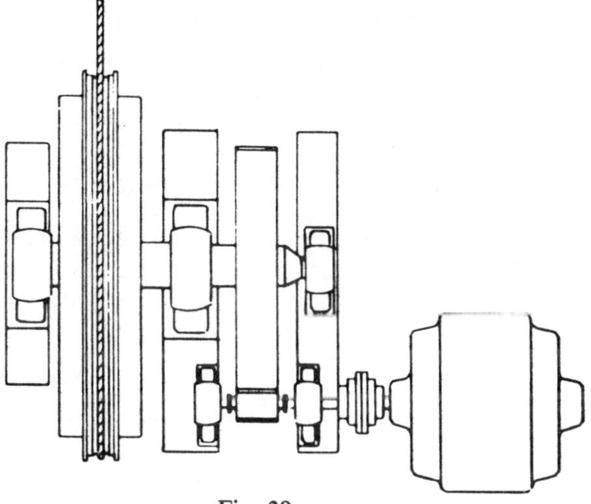

Fig. 39:
A Single Rope Friction Hoist

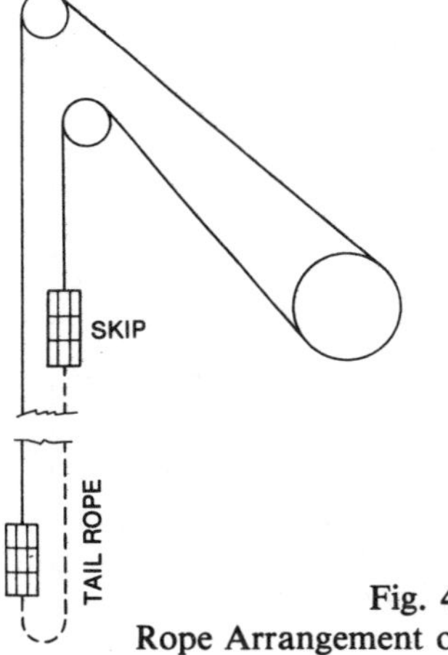

Fig. 40:
Rope Arrangement of a Friction Hoist

friction hoist, originally developed by KOEPE, using one rope as described above, is now being used with two or four such parallel ropes, each of smaller diameter. This further reduces the diameter of the drive sheaves and headframe sheaves, and therefore, the inertia losses.

In selecting a rope for a given duty, we must know the total static load on the rope when the skip is at the bottom of the shaft (see Table III). The breaking strength of the rope selected must be greater than:

$$BS = F_s (W + W_s + W_r) \text{ tons/tonnes,}$$

where W is the weight of the ore, tons/tonnes
W_s is the weight of the skip, including the draw-gear and rope attachments, tons/tonnes

Table III — Comparative Hoisting Systems

(W = weight of ore; W_s = weight of skip; W_r = weight of rope)

Balancing Status	System	Maximum static load on the rope		Net unbalanced static load on the hoist
		(a)	(b)	(a)
Unbalanced	One drum One rope One sheave One skip	$W + W_s + W_r$	—	$W + W_s + W_r$
Partially balanced	Two drums Two ropes Two sheaves Two skips	$W + W_s + W_r$	W_s	$W + W_r$
Fully balanced	*Friction winding* One drive sheave One rope One tail rope Two sheaves Two skips	$W + W_s + W_r$	$W_s + W_r$	W

(a) When the loaded skip is at the bottom of the shaft.
(b) When the unloaded skip is at the surface.

W_r is the weight of the suspended rope, from the headframe sheave to the skip at the bottom of the shaft, tons/tonnes

and F_s is the Factor of Safety, as laid down by state regulations.

There are also other considerations in the choice of rope. Hoist ropes must be regularly inspected to detect flaws or broken wires or signs of corrosion, and to measure external wear. Ropes showing significant defects must be replaced by new ropes. Splicing of a hoist rope is never allowed.

Three types of **conveyances** are used for hoisting: cages, skips, or counterweights. Except for small mines where ore cars are entered into cages for hoisting, cages are used principally for hoisting or

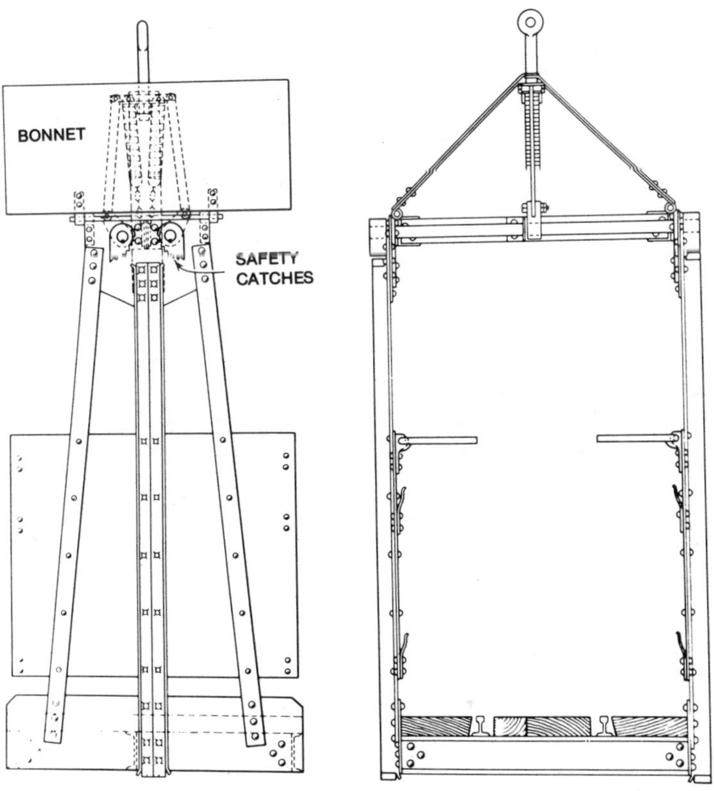

Fig. 41:
Hoisting Cage for a Small Mine

lowering men and supplies. The latter may include machinery, such as pumps, locomotives and empty ore cars; but more usually, pipes, rails, timber, explosives and general mine hardware.

From the point of view of safety, each cage must be fitted with a hood or bonnet, and with doors. Where wooden guides are used, cages are also provided with safety dogs for automatically gripping the guides and arresting the fall of a cage in case of rope breakage (Fig. 41).

In large mines, large capacity cages (to hold as many as 100 men per deck, and up to three decks) operate in balance with a counterweight in a small compartment. These cages can conveniently be loaded with large machinery items, such as locomotives, without dismantling them.

Ore skips are specially designed for the vertical transport of broken ore. Within a steel framework that runs on the shaft guides, a steel (or light metal) body carries the ore, in a few cases as much as 24 tons (tonnes). Two such skips operate in balance.

The skip is loaded at an orepocket by discharge of ore from a 'measuring flask' designed to hold a skip-load. When the skip arrives at the surface orebin, it is dumped automatically, either by overturning of the skip body about a bottom hinge; or by a bottom dump arrangement.

As each skip is loaded, the skip tender presses a button to set the hoist in motion. This is known as semi-automatic hoisting, where no winding engine driver is employed. Fully automatic hoisting is involved where the hoist is set in motion either by a pressurestat or a closed-circuit television camera.

Recommended reading:

LEWIS, R.S. and CLARK, G.B. *Elements of Mining*, 3rd edn. (New York: Wiley). 1964. pp. 239—58.

13. Mine Drainage

Some rocks are more porous than others. The rocks of the earth's crust are to a certain extent saturated below a certain level, variously called the 'water level', 'ground water level', or 'water table'. This level varies with the general climatic conditions at any particular location, and roughly follows the topographic surface level. In simple terms, it is the equilibrium level between the percolation of meteoric (rain) water from the surface, and the outflow into lakes, springs, rivers and oceans.

Although ground water naturally collects in the more porous beds, often localized by impervious strata, by and large all rocks below this level are saturated with water. This means that as shafts and other mine openings extend below this 'ground water level', water is likely to be encountered and to seep into the openings to an extent depending upon the area of rock surface exposed, the hydrostatic pressure, and other factors. In order to continue mining operations, it is therefore necessary to lower the ground water level in the vicinity of the mine by artificial means to keep the workings free of water. This operation is known as **mine drainage**.

Obviously, the first requirement is to limit the flow of surface water into the mine. Shaft collars should be set above ground level; adit portals should be provided with effective catch drains; and both should be sited above flood levels in valley situations.

In regions of high relief, where water seepage into the mines was considerable, it was formerly the practice to drive a drainage adit above river level to a point under the mine workings. This effectively lowered the ground water level and at the same time provided some opportunity for exploring the deposit. Some drainage adits have extended for up to 15 miles (24 km) in several countries.

In areas of level or undulating terrain, however, it is necessary to lower the ground water level by some mechanical means. In Roman times, a battery of water wheels, driven by slave power, progressively lifted water to the surface from the bottom of the mine. But many centuries later, steam engines were developed for mine pumping in England. This led to the Industrial Revolution in the 18th Century.

Pumps may be used externally, in boreholes situated around the periphery of the mine workings; or internally, by pumping water through a discharge pipe installed in the shaft.

A complete pumping installation consists of five main parts, viz.

(1) a collecting sump.
(2) a mechanical pump unit.
(3) a drive motor (usually electric).
(4) a suction pipe, through which water is led into the pump from the sump.
(5) a discharge pipe, through which water is delivered by the pump to the surface.

In many small mines, the bottom of the shaft, below the lowest level, serves as a **sump**, collecting the seepage water flowing by gravity along the levels. In other cases, a sump is excavated near the shaft, at one or more levels, and a pump room established nearby. The capacity of a sump is such that it will hold more than the seepage rate for one day; but provision needs to be made for future extensions to the workings. These would necessarily increase the seepage rate.

The pump needs to be located above sump level so that the motor cannot drown. The suction pipe calls for special design considerations. Pumps can be provided with self-priming features, and automatic operation. They should be of such capacity that they need to operate for say 16 hours per day, to allow a margin for maintenance time and power failures. There may be several pumps in each pump room, at least one on standby duty.

The design of the discharge pipe (rising main) is important. If of insufficient diameter, the friction head developed by the resulting high velocity flow may lead to high costs and other problems. The total head or pressure against which a pump must operate is the sum of the total static head (suction plus discharge) and the total friction head.

There are two main classes of pumps for general mine drainage duty:

(a) the reciprocating or plunger pump, operating in one or more cylinders. This is a positive displacement type of pump, under the control of suction and discharge valves. It is generally a low-speed, low-capacity, high-head pump, giving an intermittent flow, either single- or double-acting, with a single cylinder; or duplex, triplex and quintuplex models to give a more uniform flow.

(b) the centrifugal pump, consisting of a high-speed impellor rotating within a shaped volute housing, and driven by an electric motor. Each individual pump, or stage, is typically a high-speed, high-volume, low-head pump; but additional head can be gained by coupling several stages together in series. Such a multi-stage centrifugal (turbine) pump is now in regular use in large deep mines. Various designs and models of centrifugal pumps serve a range of different applications. Some of these are de-watering pumps, borehole pumps, and submersible pumps. A small portable unit (known as a sump pump), driven by compressed air, is a very useful tool.

There are many other aspects of the effects of water in mines and of the methods of dealing with them.

Recommended reading:

CUMMINS, A.B. and GIVEN, I.A., eds. *SME Mining Engineering Handbook*. (New York: AIME). Vol. 2, Ch. 26.

FREEZE, R.A. and CHERRY, J.A. *Groundwater*. (Englewood Cliffs, NJ: Prentice-Hall). 1979.

14. Mine Ventilation

All underground mines should be effectively ventilated; firstly, to provide sufficient oxygen for men to breathe; secondly, to ensure that the working climate represents comfortable conditions for men to work at maximum efficiency; and thirdly, to dilute and displace gases and dusts that would otherwise contaminate the mine atmosphere.

In the main, all these aims are achieved by passing a sufficient volume of clean fresh cool air into the workings via the main (downcast) shaft, to displace foul air exhausted through an upcast (ventilation) shaft. To cause such a flow, a pressure difference must be established to overcome the mine resistance to airflow. In small mines, this is sometimes achieved by the 'natural ventilating pressure' due to differences in the weights of the air columns in both shafts. But in most modern mines, an adequate pressure difference is secured more positively by installing an exhaust fan over the collar of the upcast shaft, reinforced by booster fans placed strategically in the main mine ventilating circuit. In all cases, the mine workings should be designed to offer minimal resistance to the flow of air.

Contaminants of Mine Air

Apart from excessive temperature and humidity problems, various harmful gases and dusts are generated, either from the rock strata or from mining operations. These include such gases as carbon dioxide, carbon monoxide, methane, nitrogen oxides, and hydrogen sulphide. Limits have been prescribed as acceptable maximum levels of concentration for each gas, either to avoid dangerous physiological problems; or to avert the formation of explosive mixtures in air, such as with methane in many coal mine atmospheres, and in some other underground situations.

When concentrations of methane occur in air between 5 and 14 per cent by volume, a violent explosion can be triggered by a spark or open flame. In practice, methane concentrations are not allowed to exceed one per cent. Methane is liberated naturally from coal seams; and carbon dioxide from certain metal mines in which limestone occurs as gangue or as wall rock. Otherwise, gases are formed from blasting operations, from decaying timber, from workmen, and from internal combustion (diesel) engines. Carbon monoxide, carbon dioxide, mixed nitrogen oxides and aldehydes are vented to the mine atmosphere by diesel engines, even though fitted with exhaust purifiers to reduce this problem.

Similarly, fine dust is generated from mining operations such as drilling, blasting and dumping; and from radioactive decay processes in uranium mines. These latter call for special precautions. Fine dust particles, generally between 0.1 and 5 microns in size, may progressively cause lung damage to miners in a range of diseases labelled as pneumoconioses.

Accordingly, all these contaminants must be expelled from the mine, in the interests of health, safety and economical working.

Temperature

In very deep mines, where rock temperatures are excessive, the ventilating air needs to be air-conditioned (refrigerated) to lower the ambient temperature and to reduce the relative humidity of the air to levels that give comfortable working conditions.

In general, rock temperatures increase by 1 °F for every hundred feet of depth (or 1 °C per 55 m). This is known as the 'geothermic gradient'.

Similarly, air temperatures in a mine increase because of adiabatic compression of the air in shafts by about 5.5 °F for every thousand feet (or 1 °C per 100 m) of depth.

Moisture control is a very important operation to reduce the level of relative humidity in a mine.

Principles of Airflow

The general expression for airflow relationships in a particular mine airway is the ATKINSON formula (expressed in imperial units).

$$H = KSQ^2/5.2A^3$$

where H is the pressure loss, inches of water
K is the coefficient of resistance, or friction factor
S is the rubbing surface of the airway (or the perimeter times the length), ft^2
Q is the air volume in units of 100,000 ft^3 per minute
A is the cross-sectional area of the airway, ft^2

The friction factor K varies widely in accordance with the degree of roughness of the particular airway lining surface.

From the ATKINSON formula, the specific resistance R of a particular airway can be deduced from the formula $H = RQ^2$, where R is equivalent to $KS/5.2A^3$. The expression $H = RQ^2$ is modified where the air density varies from the standard 0.075 pounds per ft^3. Where d is the observed density, $H = RQ^2 \times d/0.075$.

In SI units, the ATKINSON formula is expressed as

$$P = KCLv^2/A$$

where P is the pressure loss, N/m^2 (= Pa)
K is the friction factor, Ns^2/m^4
C is the perimeter, m
L is the length of airway, m
v is the air velocity, m/s
A is the cross sectional area, m^2

Since $$Q = vA \text{ in } m^3/s$$

then $$P = KCLQ^2/A^3$$

And if the specific resistance of the airway is denoted by

$$R = KCL/A^3$$

then $$P = RQ^2$$

and $$R = P/Q^2$$

The total resistance of a set of airways in series is the sum of each separate airway resistance. For airways in parallel, the resistance of a segment of parallel airways is given by

$$1/\sqrt{R} = 1/\sqrt{R_1} + 1/\sqrt{R_2} + 1/\sqrt{R_3} + \dots$$

By summing all segments in a mine ventilating circuit, the total mine resistance can be determined; and a corresponding mine characteristic curve (a parabola) can be plotted with H and Q as the axes. In selecting a main fan for a given value Q_1 of air to be circulated, the fan characteristic curve should cross the mine characteristic at a point

Q_1, H_1. At this point, the pressure head to be developed by the fan is H_1 (see Fig. 42).

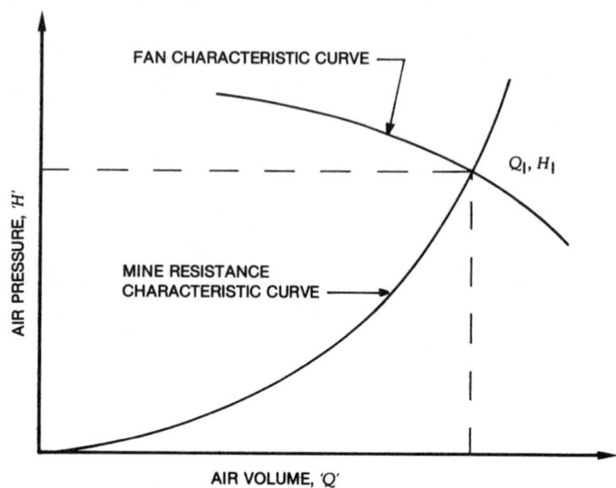

Fig. 42:
Typical Fan Selection Diagram

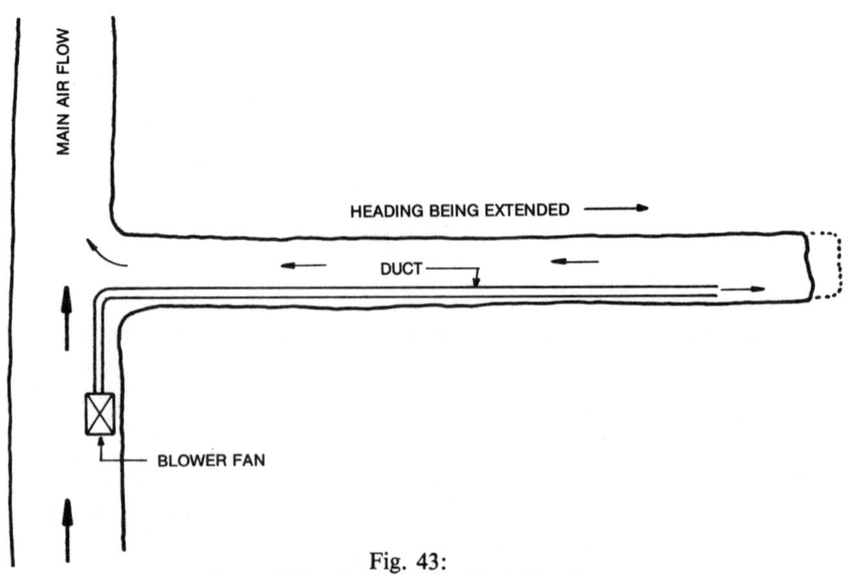

Fig. 43:
Ducted Ventilation of a Blind Heading

Secondary Ventilation

To ventilate the working faces of blind development headings, it is necessary to use a fan-duct unit. A pressure fan located in a main airway blows air through a metal or rigid plastic duct to the working face, with the return air flowing back along the development opening (see Fig. 43).

Ventilation Surveys

Routine surveys should be conducted on a regular basis to monitor the flow of air to each working face; to make observations with respect to gas and dust concentrations; and to check fan performance. All these data should be recorded and plotted on the ventilation plans. Ambient conditions in a mine change from time to time according to time of day, season of the year, and by extension of the mine workings.

A wide range of instruments is available for observing absolute pressure, pressure difference, air velocity, air temperature, air relative humidity, rock temperature, dust concentration and analysis, and the detection and concentration of specific gases.

Recommended reading:

CUMMINS, A.B. and GIVEN, I.A., eds. *SME Mining Engineering Handbook*. (New York: AIME). Vol. 1, Ch. 16.

LEROUX, W.L., *Mine Ventilation Notes for Beginners*. (Johannesburg: Mine Vent Soc of SA). 1972.

15. Other Mine Services

In addition to haulage, hoisting, drainage and ventilation, a range of other mine services must be made available to support the programme of mine development and ore production. Some of these are noted below.

Mine Lighting

The underground mine environment is basically as black as the proverbial ace of spades. Furthermore, the dark grey rock faces and black coal faces offer little or no reflectivity. It is therefore necessary to employ adequate artificial illumination in all underground mines.

In locations such as at shaft stations, pump rooms, repair shops, lunch rooms, main gangways and the like, permanent electric lighting fixtures are installed, using incandescent or fluorescent lamps or both. Reflectivity is promoted by white-washing of the walls of these areas. Where mobile equipment is employed, floodlights are fitted to the unit.

But over and above these provisions, every individual mine worker is provided with a portable electric cap lamp fitted to his helmet and powered by a battery hung on his belt. Batteries are re-charged on the surface during the off shift.

The effectiveness of artificial lighting in underground situations is adversely affected to some extent by low reflectivity of wall surfaces, by glare, and by the relative opacity of the air due to dust, fumes and water vapour. Nevertheless, poor lighting conditions must be recognized as contributing significantly to the occurrence of accidents.

In designing and surveying lighting requirements of mines, the main units used are:

(1) For the quantity of light (or the luminous flux): lumens.
(2) For the intensity of light: candelas. One candela should give 4 π lumens.
(3) For illumination (or the density of luminous flux): lumens per square foot (ft-candles); or lumens per square metre (lux).

Power and Compressed Air

Apart from mobile units, some form of power needs to be reticulated to all working faces underground for the various operations.

For drilling holes in soft material, rotary drills are applicable, and these are more readily powered by an electric motor. Therefore, in coal mines and in other mines where rocks and vein material are not especially hard, **electricity** is the normal form of energy distributed.

Armoured 3-phase high tension power cables are installed in main shafts (or through special boreholes) and extended along the main workings. At intervals, step-down transformers are installed to reduce the voltage and to correct voltage drop. Near the working faces, distribution boxes are located for service to the various working units.

In coal mines particularly, where there is a danger of triggering methane or coal-dust explosions, it is vitally necessary for all electrical equipment to be of spark-proof design.

Electric power in all underground mines is used for operating pumps, fans, trolley locomotives, scrapers and winches; and for workshop equipment, lighting, signalling, blasting, and the like.

On the other hand, where the rock is hard, such as in metal mines, the drilling of blastholes cannot be achieved economically by the rotary method. In these situations, percussive drilling must be employed; and these reciprocating drill systems cannot effectively use electric motors. **Compressed air power** is therefore used, basically to operate 'pneumatic' rockdrills. It must be reticulated throughout the mine in pipes, leading from a battery of air compressors on the surface. Also, because this network of compressed air pipes is so available, several other small portable units (such as sump pumps, auger drills, pneumatic saws and drill sharpeners) are also provided with compressed air motors.

However, for the larger stationary units, such as pumps and fans, electricity is also used in metal mines, mainly because electricity costs only about 40 per cent as much per horsepower as compressed air power.

Therefore, in these metal mines, both electric cable and compressed air networks are reticulated through the workings. In both cases, special provision needs to be made to avoid large pressure drops in long leads.

Rockdrills and Steel

For drilling blastholes in hard rocks, pneumatic rockdrills (in various designs) and tungsten carbide tipped drill steel are employed (see Fig. 22). When working effectively, these promote the productivity of individual miners; therefore they must be well maintained and provided with a consistent flow of compressed air at 90 to 100 pounds per square inch (630 to 700 kPa) pressure through hoses connected to the piped reticulation system. Similarly, a sufficient supply of sharp drill steel must be available to the drill crews to avoid lost working time. Drill bits are re-sharpened at regular intervals to maintain an effective drilling rate. Drill sharpening units are available on all main levels.

Sanitation

For reasons of health and general hygiene, chemical toilet units are normally maintained at convenient points designated for the purpose on all levels.

Similarly, well equipped lunch rooms are provided with waste collecting bins for clearance to the surface. Indiscriminate dumping of food waste underground can lead to an unhealthy environment; and of inflammable materials, to risk of fire.

General Supplies

A modern underground mine necessarily consumes a wide range of spare parts and other hardware items that must be readily available to working crews to avoid loss of working time. For this reason, stores are established on all main working levels. Typical of some of the items stocked are: pipes and pipe fittings; rails and fastenings; bolts, spikes, nails and washers; spare parts; electric cable, fittings and lamps; lubricants of all kinds; timber details; rockdrills, steel and hoses; loose tools; and explosives (in separate stores or 'magazines').

Changehouse

On all mines a changehouse is provided at the surface near the shaft. This changehouse is normally divided into three sections: (a) for miners to undress and to store clean street clothes in individual lockers, (b) a mid-section for hot showers, washbasins and toilet facilities; and (c) a section for dressing into and undressing from mine working clothes. At the end of shift, these work clothes are hoisted on hooks to allow them to dry before use on the next round of shifts.

All such changehouses are provided with ample hot and cold water, and an air-heated section for underground clothes. They are maintained by a regular attendant.

In cold climates, a protected subway tunnel leads from the changehouse to the shaft brace where men are loaded into the cage, and vice-versa.

Recommended reading:

TROTTER, D., *The Lighting of Underground Mines*. (Clausthal-Zellerfeld: Trans Tech Publications). 1982.

16. Mineral Processing

As mentioned in Chapter 3, most run-of-mine materials need some form of beneficiation to remove waste matter and therefore to upgrade the valuable product to meet market conditions. These beneficiation plants are variously called mills, treatment plants, concentrators, coal washeries, or coal washing plants. They are usually established on the surface near the main shaft or adit portal.

Much of the waste material unavoidably mined along with the useful mineral consists of wall rock, roof rock or floor rock; but also of bands of waste or 'dirt' within the vein, lode or seam. This material is fairly readily removed in the beneficiation process.

However, in the case of metalliferous veins or lodes, the vast proportion of waste consists of gangue minerals, representing the matrix of the vein. See Chapter 3. The useful minerals are irregularly dispersed and intimately mixed as fine specks, blebs or veinlets within this matrix or groundmass. The separation of these useful minerals from the gangue minerals therefore presents a major problem. In most cases, the run-of-mine material must be crushed and ground into very fine particles to separate them physically.

Therefore, the first stage of beneficiation of metalliferous run-of-mine ore is to crush and screen it in various stages to about 1 1/2 inches (4 cm) in size, in jaw crushers or gyratory crushers. Water is then added as the crushed ore is fed to a battery of ball mills to grind it to a particle size sufficient to liberate the useful minerals from the gangue. The final particle size is controlled by some form of classifier unit. The undersize portion is then ready to pass to the concentration process, which depends upon the type and character of the useful minerals in the ore.

There are several different classical methods of ore concentration, viz.,

By differences in specific gravity.
By differences in magnetic susceptibility.
By differences in electrostatic conductivity.
By different responses to various chemical reagents in an agitated pulp (the 'flotation process').
By dissolution in sodium cyanide solution (as for gold and silver).
By alloying with mercury (the 'amalgamation process').
By dissolution in various leaching solutions.
By other methods.

Several of these methods may be used in combination to treat any one type of ore, especially where a number of products can be recovered.

The most commonly used method of beneficiating low-grade sulphide ores is the flotation process of concentration. The pulp (of finely ground ore in water) is aerated and agitated while certain chemical reagents are added. These reagents cause the surfaces of the desired mineral particles to be lifted by air bubbles to the surface of the pulp, where they are skimmed off as a frothy concentrate. The gangue minerals do not respond to the reagents and therefore remain in the pulp from which they are filtered and despatched to the dump as tailings. Meanwhile, the frothy sulphide mineral concentrate is filtered, dewatered or dried, and despatched to the smelter for recovery of the metal in ingot form for sale in the market.

Other methods of concentration and metallurgical smelting have been used and applied specifically for each family of minerals. Each method involves the disposal of the worthless gangue and other waste materials as tailings.

Exceedingly close controls are employed to ensure that the efficiency of recovery of the minerals and the grades of the concentrates produced are maintained at a high economic level.

Recommended reading:

Cummins, A.B. and Given, I.A., eds. *SME Mining Engineering Handbook*. (New York: AIME). 1973, Vol. 2. Ch. 27.

17. Mine Management

For effective control and successful management of all aspects and operations of a modern mine, the General Manager needs to have outstanding training, experience and personal qualifications, with a broad outlook, a flair for administration and demonstrated powers of leadership. Similarly, his wife should be of such character and bearing that with gracious sincerity she commands the respect of the womenfolk, and indeed the whole mine community.

Ideally, the General Manager will be qualified with a baccalaureate degree in mining engineering, perhaps supported by an MBA degree, and with basic practical experience gained through the **operational** (production) side. He will make a much better general manager than a mining engineer (or for that matter, any other engineer) with a Ph.D. degree, earned through experience gained only through the **technical** side of mining.

Desirable personal qualifications for general managership should include significant powers of verbal exposition and chairmanship, and proficiency in such social achievements as bridge, golf, and the ability to "hold his liquor". These are important because the general manager and his wife will naturally lead the social life and promote the morale of the mining camp, and at times they will need to be hosts to visiting dignitaries.

Every young mining engineer during his baccalaureate programme should bear the above points in mind, because later in his career he could become a candidate for a general managership.

Answerable to the General Manager, will be the heads of each department of the mine operations, such as the Mine Superintendent, the Mill Superintendent, the Smelter Superintendent, the Chief

Engineer and the Chief Accountant. Therefore, he will need to have a broad knowledge and interest in all these fields for proper control of their activities.

If the mine is based on a remote locality, the management needs to establish a range of support facilities known as 'infrastructure'. This may include a powerhouse, water supply, employee housing, amenities, health facilities, schools, communications of all kinds, shopping and catering services, workshops, and warehouses. See Chapter 7. Many of these are necessary to maintain morale and to reduce labour turnover. The cost of recruiting and training new employees at short recurrent intervals for a remote mining operation can be prodigious.

Some of the important features of mine management are listed below.

Staffing

For a large mine, apart from an adequate contingent of mining engineers, the salaried staff would consist of such professionals as geologists; metallurgists; mechanical, civil, electricial and industrial engineers, draughtsmen, surveyors, assayers and the like; as well as such support staff as accountants, medical officers, environmentalists, lawyers and economists; and supervisory personnel such as foremen and shift bosses.

Industrial Relations

Miners are known to constitute one of the most rugged and fractious sectors of the national work force, based upon a tradition of manliness, tough working conditions and pride. As such they can be difficult as a body to control, especially when new methods are to be introduced.

However, there are usually no particular problems where there is an enlightened management with an innate understanding of the mental processes of miners. Miners expect the management to be firm but fair in all dealings. If management is weak or vacillating, or if a promise is not honoured, then the management will lose the confidence of the mine workmen, and from then on it is suspect. Such a situation can lead in time to industrial anarchy.

Within the mine industrial office, hiring, firing, payroll activities and all forms of statistics related thereto are carried out. One of the commonest causes of loss of confidence in management derives from doubts and suspicions about payroll calculations. To avoid this, it is good practice to seal an employee's net cash pay in a specially printed envelope designed to show on the outside the amount of his gross pay, each separate deduction, and his net pay, which must obviously correspond with the amount of cash in the envelope.

Most legal labour contracts make special provision for penalty surcharges such as overtime, late shifts, Saturday, Sunday and holiday work. If management tries to evade any of these provisions without a legal approach, it will lose face, and faith.

For continuous shift work operations (24 hours per day, seven days per week), as in the mill and the smelter, men rotate shifts on a weekly basis. For this 168 hours per week, four crews of men can be rostered to work an average 40 hours per week (or whatever) with a minimum of overtime (see Table IV).

Environmental

In some countries over a number of past years, and especially in highly vegetated areas, a great deal of environmental damage was created by some mining companies by indiscriminately clearing trees, disturbing the land surface, and emitting corrosive gases from smelter stacks. Most of these companies are now defunct; but the recent welter of public criticism in the United States has pointed the finger of scorn, somewhat unfairly, at *all* mining operations.

These critical outbursts, some of which are well-founded and others fanatical, have at least drawn attention to the need for better environmental practices by mining and other industrial operators generally.

Nevertheless, many mining companies in several countries have always taken pride in maintaining pleasant environmental conditions around mining sites, even in desert areas, as a measure of mine community planning, even at considerable expense. Such practices should be encouraged.

Health and Safety

Depending upon the size of the mine, it may be necessary to set up a hospital for all employees and their dependents; or a clinic, or first aid station for treatment of industrial injuries.

Coupled with these facilities, management should sponsor and support a genuine interest in 'safety first' procedures. This can be done by safety posters, by competitive safe working records of crews and by prize awards.

Table IV — Continuous Shift Roster

For four crews (A, B, C, D) working five 8-hour shifts (plus one overtime shift for one crew) per week in rotation. Total hours per week are 168. No 'quick' shifts. Roster is repeated every three weeks.

	Shift	Sun	Mon	Tue	Wed	Thur	Fri	Sat
First week	Night 12—8	C	C	D	D	D	D	D
First week	Day 8—4	A	A	A	A	A	*	B
First week	Afternoon 4—12	B	B	B	B	C	C	C
Second week	Night	A	A	A	A	A	B	B
Second week	Day	B	B	B	*	C	C	C
Second week	Afternoon	C	C	D	D	D	D	D
Third week	Night	B	B	B	C	C	C	C
Third week	Day	C	*	D	D	D	D	D
Third week	Afternoon	A	A	A	A	A	B	B

* Overtime shift shared evenly by each crew in turn.

Cost Accounting

All management excecutives should take a healthy interest in cost accounting procedures, in addition to the normal routine accounting records.

To counter inflationary trends and to cope with the increasing costs of operation as a mine gets deeper, there is continual pressure to keep working costs within acceptable limits. Such a cost-cutting programme is exceedingly difficult unless reliable unit working costs are available, period by period, for individual operating sectors. Again, such reliability depends upon the adoption of an effective array of cost accounts, and upon the faithful allocation of charges to each such account by means of the usual labour cards and stores requisitions. This requires the active co-operation of supervisory staff at all levels.

General

There is tremendous scope in mining for the use of industrial techniques such as sophisticated scheduling procedures and aids to decision-making and optimization. Every mining engineering school should include in its programme a required course in Operations Research (Systems Analysis).

One of the outwardly simple aspects of any industrial operation is 'good housekeeping'. But this apparently minor feature is much more important than it seems.

By continuing to avoid the accumulation of dirt, dust, waste papers, and unused scrap materials of all kinds in a generally untidy manner all around the mine (surface and underground, in buildings and in the open), a greater degree of respect for efficient working can be engendered in all ranks, and safer working conditions can ensue.

Men who work in tidy situations have a greater respect for the organization generally; they are much more likely to take pride in the quality and quantity of their work output; and they are much less inclined to cause industrial trouble.

In the whole gamut of industrial operations in western economies, there is no more complex organization than that of a modern metalliferous underground hardrock mine. This is dictated by the natural vagaries of the orebody, the continuing problems of drainage, ventilation and hoisting as the mine becomes deeper, the rugged character of mine workmen, the problems of working in remote locations, the continuing search for more ore as ore extraction proceeds, the inflationary pressure of new wage demands, the worldwide shortage of trained mining engineers, the volatility of metal prices on world markets, and the unconscionable greed of governments for higher taxes and royalties.

The manager who can cope with these problems successfully is fitted by such administrative and industrial experience to run any other organization, whatever its nature.

This is what makes mining so interesting and fascinating. What a challenging opportunity for the young mining engineer!

Appendix

Conversion Table of Units

SI (Metric) Units	Imperial Units		Reciprocals
Length			
millimetre (mm)	0.03937	inches (in)	25.40
centimetre (cm)	0.3937	inches	2.54
metre (m)	3.281	feet (ft)	0.3048
kilometre (km)	0.6214	miles	1.609
Area			
square centimetre (cm^2)	0.1550	in^2	6.452
square metre (m^2)	10.76	ft^2	0.0929
Volume			
cubic centimetre (cm^3)	0.06	in^3	16.39
cubic metre (m^3)	1.308	yd^3	0.7646
Mass			
gramme (g)	0.03215	troy ounce	31.10
kilogramme (kg)	2.205	lb avoirdupois	0.4536
tonne (metric ton = 1,000 kg)	1.102	sh tons (2000 lb)	0.9072
tonne	0.9842	lg tons (2240 lb)	1.016
Miscellaneous			
gramme/tonne	0.029	troy oz/sh ton	34.28
kg cal (4185 J)	3.968	BTU	0.2520
cal/g	1.8	BTU/lb	0.56
kg/m^2 (9.81 Pa)	0.001422	lb/in^2	702.97
Pa = N/m^2			

Glossary of Mining Terms

1. **ADIT.** A blind horizontal opening into a mountain, with only one entrance.
2. **ASSAY.** The testing of a sample of minerals or ore to determine the content of valuable minerals in the sample.
3. **BACK.** The ceiling of any underground excavation in a metalliferous mine.
4. **BALL MILL.** An item of milling equipment used to grind ore into small particles. It uses steel balls as the grinding medium.
5. **BASE METAL.** A commercial metal such as copper, lead or zinc. The term is used to distinguish these metals from the precious metals.
6. **BEDROCK.** The solid rock of the earth's crust, generally covered by overburden, or gravel, soil, or water.
7. **BEDDED DEPOSIT.** A mineral deposit of tabular form that generally lies horizontally and is commonly parallel to the stratification of the enclosing rocks. A coal seam is a typical bedded deposit. Others may contain industrial minerals, and some are metalliferous.
8. **BENEFICIATE.** To treat mined mineral matter so that the resulting product is richer or more concentrated with the useful mineral.
9. **CAGE.** An elevator-type conveyance which moves men and materials up and down a mine shaft by means of a hoist rope.
10. **CALORIFIC VALUE.** The quantity of heat units that can be liberated from one pound of coal or oil, when measured in British Thermal Units (BTU). Coal, being a 'mineral' of organic origin with no precise composition, cannot conveniently be evaluated in terms of grade. It is therefore evaluated in terms of its calorific value (CV).

11. **CHUTE.** An opening in mine workings through which broken ore is moved into mine cars, for haulage to the shaft.
12. **CLAIM.** An area of land staked by a prospector or mining company and then recorded (see Staking).
13. **COAL RESERVES.** Measured tonnages of coal that have been calculated to occur in a coal seam within a particular property.
14. **COLLAR.** The entrance-way from the surface, of a shaft; or from a level, of a winze.
15. **CONCENTRATE.** To treat ore so that the resulting 'concentrates' will contain less waste and a higher amount of valuable mineral. In many mining operations, ore is concentrated in a 'concentrator' or 'mill' on the surface, and then shipped to a smelter for the recovery of metal (see Beneficiate).
16. **CORE.** A cylindrical stem of rock that is extracted from the earth by a diamond drill. The core is removed to the surface for examination and/or analysis (assay).
17. **CROSSCUT.** A horizontal opening driven across the strike of a vein, or across the direction of the main workings. A connection from a shaft to the vein.
18. **CUT-AND-FILL STOPE.** A stope in which the ore is removed in slices, after which waste material (backfill) is run in before the next slice is mined. The backfill supports the walls of the stope.
19. **DEVELOPMENT.** The work of driving openings to and in a proved orebody, or a coal seam, to prepare it for systematic mineral production.
20. **DIP.** The angle at which the vein is inclined from the horizontal.
21. **DRIFT.** A horizontal opening in or near an orebody, parallel to the strike of the vein, or the long dimension of it. Otherwise, a type of coal mining in hilly country where the seam is entered horizontally (without the use of a shaft).
22. **EVALUATION.** The work involved in gaining a knowledge of the grade, shape, position and size of a prospect.
23. **EXPLOITATION.** The work of mining and processing the ore, for sale in the market.
24. **EXPLORATION.** The search for a mineral deposit (prospecting) and the subsequent investigation of any deposit found (evaluation).
25. **FAULT.** A break in the earth's crust caused by forces which have moved the rock on one side with respect to the other.

26. **FERROUS.** Minerals that contain iron. Minerals of metals that do not contain iron are termed 'non-ferrous'.
27. **FILL** (or Backfill). Waste material used to support the walls of a stope, and to provide a working platform for the miners.
28. **FLOTATION.** A commonly used milling process in which certain minerals in a water-borne pulp attach themselves to air bubbles and float to the surface, while gangue minerals sink to the bottom or pass out the exit, thereby causing separation.
29. **FOOTWALL.** The wall of rock under an inclined vein. It is called the 'floor' in bedded deposits.
30. **GANGUE.** The mineral material in an ore, forming part of a vein or lode, that is not commercially useful. Gangue minerals are discarded as tailings as soon as they can be separated from the useful or valuable minerals, during the concentration process.
31. **GRADE.** The proportion of metal contained in unit weight of ore, even though the metal is in the form of a mineral. The grade is usually expressed as a percentage by weight. For instance, a grade of 4% lead means that a short ton (2000 lb) of lead ore contains 80 lb of lead, usually in the form of the mineral 'galena' (PbS). In the metric (SI) system, a grade of 4% lead means that a tonne of lead ore contains 40 kg of lead. The ores of precious metals contain minor amounts of the metal and cannot conveniently have their grades expressed in percentages. They are therefore expressed as troy ounces per short/long ton; or as grammes per tonne.
32. **HANGING WALL.** The wall of rock on the upper side of an inclined vein. It is called the 'roof' in bedded deposits.
33. **HAULAGE.** The horizontal transport of broken ore along a level to an orepocket near the shaft.
34. **HOISTING.** The vertical transport of broken ore up the shaft from the orepocket to the orebins on the surface.
35. **INDUSTRIAL MINERALS.** Usually non-metallic minerals which are used in industry and manufacturing processes in their natural state, though generally with some beneficiation to imposed specifications. Examples include asbestos, salt, potash, trona, and phosphate rock.
36. **JAW CRUSHER.** A machine in which the ore is reduced in size by the action of moving steel jaws.

37. **LEVEL.** Mines are customarily worked from vertical shafts through horizontal passages (drifts and crosscuts) called 'levels'. These are commonly spaced at regular intervals in depth, and are either numbered from the surface in regular sequence, or designated by their actual distance below the collar of the shaft.
38. **LODE.** A wide near-vertical mineral deposit in a sheared zone of rock.
39. **MILL.** A plant, usually at the mine site, which concentrates ore or treats it so that minerals are separated and prepared for ultimate recovery in a purer form.
40. **MINERAL.** An inorganic compound occurring naturally in the earth's crust, with a distinctive set of physical properties and a definite chemical composition. Minerals are therefore assemblages of chemical elements. They may be described as 'common rock-forming' minerals, or as minerals 'of economic value'. For convenience, native metals and organic hydrocarbon materials (like coal and petroleum) are classified as minerals.
41. **MINING.** The process of obtaining useful minerals from the earth's crust for the benefit of mankind. Mining = mineral production. It includes both underground excavations and surface workings.
42. **MUCK.** Ore or rock that has been broken by blasting.
43. **OPTION.** A right to have the first chance to buy or refuse to buy a mining property.
44. **ORE.** A natural aggregate of metalliferous minerals that can be extracted from the earth's crust at a profit. Ore contains commercially useful minerals as well as 'gangue' minerals.
45. **OREBODY.** A mineral deposit containing ore, such as a vein or lode.
46. **ORE DEPOSIT.** A metalliferous deposit sufficiently concentrated by nature to warrant extraction by mining.
47. **ORE POCKET.** An excavation in the rock near the shaft to store broken ore delivered by haulage trains, with chute gates feeding skips for hoisting to the surface.
48. **ORE RESERVES.** Measured tonnages of ore of a certain grade in a particular deposit. May be described as proven, probable or possible, depending upon the accuracy with which the size and grade of the ore has been determined.

49. **OUTCROP.** The surface exposure of a mineral deposit. In some cases, where the uppermost part of a deposit is concealed by soil or overburden, there may be no outcrop.

50. **OVERBURDEN.** Soil, plant life, or water that covers solid rock, or a mineral deposit.

51. **PORTAL.** The entrance to a tunnel or adit.

52. **PROSPECT.** A mineral deposit, the value of which has not adequately been tested or evaluated.

53. **PROSPECTING.** The search for a mineral deposit in the earth's crust.

54. **RAISE.** A vertical or inclined opening driven upward from a level to connect with the level above, or to explore the ground for a limited distance above a level.

55. **ROCK.** An assemblage of common minerals usually in random proportions (with no definite chemical composition). Usually, any useful minerals present in a rock are too dispersed to be of commercial value. But some rocks themselves are commercially useful, such as building stone, monumental stone, aggregate for concrete, etc.

56. **ROYALTY.** Amounts of money paid by an operating mining company to the actual owner of the mineral rights to the property. The royalty may be based upon an agreed amount per ton, or a percentage of the revenue or profits.

57. **RUN-OF-MINE.** Ore of average grade in a mine.

58. **SHAFT.** A major development opening. A vertical (or inclined) excavation in a mine extending downward from the surface, or from some interior point, as a principal opening through which the mine is exploited. A shaft may be provided with a hoisting engine and headframe at the top for handling ore, men and supplies; or used only in connection with pumping or ventilating operations, or to provide an escapeway. A shaft generally is divided into separate compartments.

59. **SHRINKAGE STOPE.** A method of stoping that utilizes part of the broken ore as a working platform and as temporary support for the walls.

60. **SMELTING.** The recovery of metal ingots from ore which has been treated and concentrated at a mill; smelting is required to recover the metal content in a form from which it can be sold or further refined before sale.

61. **STAKING.** The measuring and marking with stakes or posts of an area on the ground to establish mineral rights.
62. **STOPE.** The part of an orebody from which ore is currently being mined, or broken, by stoping (drilling and blasting).
63. **STRIKE.** The horizontal course or bearing of an orebody, such as a vein. The direction of a horizontal line in the plane of the vein. The direction of a vein as it would appear on a horizontal land surface.
64. **SUMP.** An excavation for the purpose of collecting or storing seepage water before pumping to the surface. The bottom of a shaft is sometimes used for this purpose.
65. **TUNNEL.** A horizontal opening clear through a mountain, with two entrances.
66. **VEIN.** A mineral deposit having a more or less regular development in length, width, or depth to give it a tabular form, commonly inclined at a considerable angle to the horizontal. The term 'lode' is similar to a vein, but of much greater width, and with other characteristics.
67. **WALL ROCK.** The rock forming the walls of a vein or lode.
68. **WASTE.** Material that is too low in grade to be of economic value. Broken barren rock. Mullock.
69. **WINZE.** A vertical or inclined opening sunk from a point inside a mine for the purpose of connecting with a lower level, or of exploring the ground for a limited depth below a level.

INDEX

Index

Abrasives, 23
Acknowledgements, 9
Adit, 41, 42, 52, 90, 103, 117, 129
Aerial photography, 34
Alimak unit, 42, 43
Aluminium, 16, 23
Alluvial mining, 40, 62
Amalgamation process, 63, 120
Anomaly, 34
Anthracite, 23, 32, 40
Antimony, 27
Ash, 30—32
Assaying, 33, 35, 37, 38, 129
Arizona, 57
Atkinson formula, 110, 111
Auger mining (of coal), 40, 56, 64
Australia, 55, 63
Austria, 65

Barite, 23, 31
Base metals, 23, 38, 129
Bedded deposit 25, 26, 29—31, 34, 39—41, 51, 56, 57, 67, 70, 80, 96, 129
Bench, 39, 61
Beneficiation, 29, 32, 69, 119, 129
Blasting agent, 76, 77
Blasting, powder, 76
Blasting with explosives, 48, 50, 56, 57, 61, 68, 73, 76, 77, 116
Breaking ground, 69, 73, 76
Breaking strength (of a rope), 102, 103
Britain, Great, 64, 106
Bucket wheel excavator, 56, 59
Bunton 41, 42

Cage, 42, 43, 45, 90, 95, 97, 103, 104, 118, 129
California, 62
Calorific value, 31, 32, 38, 129
Changehouse, 118
Chromium, 27
Chute, 47, 48, 50, 51, 92, 130
Chute gate, 47, 48, 50, 90, 92, 93
Chute timbering, 50, 51, 80, 92
Coal, 17, 20, 23, 25, 30—32, 34, 40, 51, 53—56, 64, 87, 90, 93, 96
Coal mining, 26, 51, 70
surface, 56
underground, 52—56, 109, 116
Coal seam, 25, 51—53, 56, 64, 110, 119
Collar, 45, 97, 104, 130
Compartment, 41, 45, 46, 104
Compressed air, 64, 65, 73, 116, 118
Concentration (*see* Beneficiation)
Concentrate, 29—31, 67, 69, 120, 130
Concentrator (*see* Mill)
Continuous miner, 53, 55, 56, 93
Copper, 16, 17, 23, 27, 57, 63, 65
Conversion table, 127
Conveyance, 125
Conveyor belt, 53—55, 87, 90, 93, 96
Core, 35, 37, 130
Cost accounting, 125
Costs (of mining), 24, 26, 27, 39, 42, 46, 56, 61, 67, 68, 101, 117 125
Crosscut, 42, 45, 46, 53, 54, 92, 130
Crown pillar, 48—51
Cut-and-fill mining, 27, 49, 51, 130
Cut off grade, 26—28
Cyanidation process, 120

Detonator, 76, 77
Development, mine, 39, 52—54, 61, 67—69, 87, 90, 130
Diamond, 17, 23, 62
Diamond drill, 34, 35, 37, 45, 67
Dragline excavator, 56—58
Drainage adit, 105
Drawbar pull, 94
Drawpoint, 87, 88, 90, 92
Dredging, 62
Drift, 42, 45, 46, 52, 72, 80, 92, 130
Drilling, 34, 51, 57, 61, 64, 68, 73, 76, 77, 116
Dumping, 61, 87, 90, 92, 95, 104
Dust, 31, 47, 53, 73, 107, 110, 113, 125
Dynamite, 76

Economic mineral, 13, 20, 23, 24, 26, 28, 29, 39
Electronic industry mineral, 23
Electric power in mines, 31, 67, 68, 93, 95, 116, 124
Environmental, 46, 119, 123, 125
Evaluation (of a mineral deposit), 35, 37, 38, 67, 130
Explosions in coal mines, 109, 116
Explosives for blasting, 26, 29, 46—48, 53, 73, 76, 77, 104, 117

Fan selection diagram, 112
Feasibility report, 36, 67, 68
Financial aspects of mining, 36, 68, 69
Floor rock, 52, 119
Flotation process, 120, 131
Footwall, 41, 45, 47—49, 61, 85, 131
Frasch process, 40, 65
Friction hoist, 99, 102
Fuel minerals, 23, 29—31, 40

Galena, 19
Gangue, 29, 30, 69, 110, 119, 120, 131
Geiger-Mueller Counter, 34
Gemstones, 23
Geochemical prospecting, 33, 34
Geothermic gradient, 112
Glossary of mining terms, 24, 26, 42, 129—134
Gob (goaf), 55
Gold, 16, 17, 23, 27, 38, 62, 63
Grade (of ore), 21, 24, 25—27, 32, 34, 35, 39, 56, 67, 131
Great Salt Lake (Utah), 62
Grizzly, 90
Gunite, 79, 84

Hanging wall, 45, 47—49, 61, 85, 131
Haulage (underground), 39, 42, 47, 53, 68, 87, 92, 93, 115, 131
Headframe, 45, 52, 95, 101
Hoist rope, 45, 97—104
Hoisting, 42, 45, 48, 51, 52, 68, 87, 90, 97—104, 115, 125, 131
Housekeeping, 125
Hydraulic mining, 63

Illumination in mines (*see* Lighting)
Indonesia 63
Industrial minerals, 23, 25, 29—31, 34, 40, 56, 57, 73, 131
Industrial relations, 122, 123, 125
Industrial Revolution, The, 102
Infrastructure, 68, 70, 122
Ingots, 30, 31, 69, 120
Insulating materials, 23
Iron, 16, 23, 25, 27, 31

Koepe hoist, 102
Lake deposits, 40
Lead, 16, 17, 23, 27, 31
Level, 42, 45, 47—51, 68, 71, 72, 90, 92, 93, 106, 133
LHD units, 87, 90, 92
Light metals, 23
Lighting (of a mine), 68, 115, 116

Lignite, 23, 25, 32
Limestone, 18, 20, 23, 25, 110
Loading, 51, 53, 57, 61, 64, 68, 87, 88, 90, 92, 93
Locomotive, 87, 93, 94,96, 104, 116
Lode, 20, 21, 25, 28—30, 33, 34, 39, 51, 56, 61, 67, 119, 132
Longwall mining, 54—56, 87

Malaysia, 63
Manganese, 23, 27
Marble, 20
Mendelèef's Periodic Table, 15
Meridian, transfer of, 71, 72
Metal prices, 24, 26, 27, 36, 68, 126
Metalliferous minerals, 23, 25—27, 29—31, 40, 46, 56, 57, 61, 67, 69, 73, 116, 119
Methane, 109, 110, 116
Mill, 29, 30, 61, 63, 68, 69, 119, 132
Mine development, 39, 46, 47, 52—54, 115
Mine drainage, 39, 42, 51, 53, 68, 105—108, 115, 125
Mine fan, 46, 110, 112, 116
Mine management, 67, 121—126
Mine support, 49, 54, 55, 68, 69, 79—86
Mineral, 13, 15—18, 20, 24, 26, 27, 29, 32, 34, 37, 46, 62—64, 120, 132
Mineral deposit, 20, 21, 23, 24, 28, 33, 35, 36, 39, 56, 62, 65, 67
Mineral exploration, 33, 34, 62, 130
Mineral processing, 29, 119, 120
Mineral production, 68, 69
impact of, 13—15
importance of, 14, 15
Mining industry, 24,
Mining terms (Glossary), 24, 26, 42, 129—134
Mining, what is? 13
Modes of occurrence (of mineral deposits), 24
Molybdenum, 27
Native metals, 17
New Mexico, 57
Nickel, 23, 27, 63
Nitroglycerine explosives, 76
Non-entry mining, 40, 64
Non-metallic minerals (*see* Industrial minerals)

Ocean mining, 40, 63
Oil wells, 40, 64
Open cut work, 26, 27, 40, 57, 61, 77
Operations research, 130
Ore cars, 47, 48, 87, 88, 90—93, 96, 103, 104
Ore reserves, 35, 36, 39, 134
Ore skip (*see* Skip)
Orebody, 27—29, 34, 35, 37, 39, 40, 51, 65, 67, 69, 132
Outcrop, 33, 34, 133
Overburden, 39, 56, 57, 61, 64, 73, 133

Payroll, 122, 123
Peat, 25, 32
Petroleum, 17, 23, 64
Pillars, 53, 54, 56, 80
Placer mining (*see* Alluvial mining)
Platinum, 17, 23, 27, 62
Pneumoconioses, 110
Power shovel, 56, 57, 60, 61
Powered supports, 50, 55
Precious metals, 23
Prospecting, 33, 35, 67, 133
Pumps, 102, 106, 107, 116
Pyrite, 27, 31

Quarrying, 40, 61, 77

Radio-active metals, 23, 34
Raises, 42, 46, 135
Replacement deposits, 25, 27
Resistance to airflow, 42
Restoration 69

Re-vegetation, 56
Rock, 15, 18—21, 24, 25, 29, 33, 35, 41, 42, 45, 46, 48, 53, 61, 62, 73, 79, 84, 105, 116, 130
Rock breaking, 69, 73, 76
Rock mechanics, 79—86
Rock pressure 42, 51, 79
Rock temperature, 51, 113
Rockbolts, 54, 80, 84, 85
Rockdrills, 47, 61, 73—76, 116, 117
Rocker shovel, 88, 90
Roof rock, 52, 53, 55, 56, 72, 119
Room-and-pillar mining, 53—56, 80, 87, 93
Royalty, 133
Rutile, 62

Safety, 39, 46, 61, 93, 103, 104, 112, 123, 125
Safety fuse, 76, 77
Salt mining, 62—65
Sampling, 32—35, 37, 38, 68
Sanitation, 117
Saudi-Sudanese
  Red Sea Commission, 63
Scintillometer, 34
Scram drift, 87, 90, 92
Scraper unit, 51, 62, 87, 89, 90, 92, 93, 116
Secondary ventilation, 112, 114
Shaft, 41, 42, 45—48, 52, 53, 68, 71, 72, 84, 87, 90, 93, 95, 97, 102, 105, 106, 109, 133
Shaft timbering, 41, 42, 80
Shift roster, 123, 124
Shortwall mining, 55, 56
Shotcrete, 79, 84
Shrinkage, stope, 27, 48, 133
Shuttle car, 87, 89, 90, 93
Siberia, 62
Silver, 17, 23, 27, 38, 63
Skip, 41, 42, 45, 48, 90, 95, 101, 102, 104
Slag, smelter, 31, 67
Slope angle (safe), 61
Sluice box, 62
Smelter, 26, 30, 31, 67, 69, 120, 123, 133
Solution mining, 40, 65
South Africa, 51
Soviet Union, 64
Square set, 26, 27, 80—82
Steel industry metals, 23
Stope filling, 49—51, 69, 80, 85, 92, 131
Stoping, 46, 47, 67, 68, 85, 134
  cut-and-fill, 49—51, 85, 92, 132
  other, 51
  shrinkage, 48, 49, 133
Strip mining, 40, 56
Stull, 48, 49, 80
Strike, 25, 33, 46, 134
Sulphur, 17, 23, 40, 65
Sump, 45, 108, 134
Surface mining, 26, 28, 39, 40, 56, 57, 61, 62, 67, 73
Surveying, 68, 71, 72
Swedish train, 95, 96

Tailing, 29, 30, 69, 80, 85, 120
Temperature (of air), 110, 113
Temperature (of rock), 110, 113
Thailand, 63
Timbering, 41, 42, 48, 51, 54, 80, 81
Tin, 23, 27, 62, 63
Topsoil, 56, 57
Tractive effort, 94
Transportation, 61, 64, 67, 87, 90, 93, 94
Tungsten, 23

Underground gasification (of coal), 40, 64
Underground mining, 25, 28, 39, 40, 45, 61, 62, 64, 69—71, 92, 109, 116, 125

United States (of America), 25, 64, 123
Uranium, 23, 65

Vanadium, 23
Vein, 20, 21, 24, 25, 29, 31, 33—35, 37, 39 41, 42, 45, 48, 62, 87, 119, 134
Vein mining, 40, 45, 47, 51, 67
Ventilation, 39, 42, 46, 47, 51—53, 68, 92, 109—115, 127

Wagner scooptram, 88, 92
Whim, 97, 98
Whip, 97, 98
Windlass, 97, 98
Winding engine, 45, 97, 104
Winze, 42, 136
Wire rope, 41, 45, 92, 95, 97—101

Yieldable steel sets, 80, 83, 84

Zinc, 23, 27, 63
Zircon, 23

About the Author

Cedric E. Gregory was born in Adelaide, Australia and received his first degree in mining engineering at the University of Adelaide during the Depression. Subsequently, he deliberately sought practical employment as an underground hardrock miner to prepare himself for professional life. Soon after, he became a senior executive in the management of mining operations. Following four years of World War II duty with the Royal Australian Engineers in the Pacific, he entered industry as a secondary career and occupied senior positions in industrial management until he suffered a severe accident which required two years of recovery.

After his convalescence, he entered academic life as a university instructor. This proved most satisfying because his extensive background of practical mining experience brought to the classroom a rich supplement to textbook theory.

Gregory worked hard to enhance his academic status, devoting six years of personal research to a crucial aspect of mine ventilation for which he subsequently earned an international reputation.

Dr. Gregory retired as Professor Emeritus of Mining Engineering from the University of Idaho in 1974, and made preparations to live quietly on the Mediterranean coast of Spain. However, he has since been called out as a Visiting Professor at Virginia Polytechnic Institute and State University and at King Abdulaziz University in Saudi Arabia.

His major hobbies are world travel, writing, inspiring students, and attending international conferences, activities in which he is ably and devotedly assisted by his charming wife, as it has been throughout their 48 years of union. They especially enjoy periodic visits to Australia to visit with their five children and 15 grandchildren.